Indigenous Peoples and Politics

Edited by:
Franke Wilmer
Montana State University

A Routledge Series

Indigenous Peoples and Politics

Franke Wilmer, General Editor

Inventing Indigenous Knowledge
Archaeology, Rural Development, and the Raised Field Rehabilitation Project in Bolivia
Lynn Swartley

The Globalization of Contentious Politics
The Amazonian Indigenous Rights Movement
Pamela L. Martin

Cultural Intermarriage in Southern Appalachia
Cherokee Elements in Four Selected Novels by Lee Smith
Katerina Prajnerova

Storied Voices in Native American Texts
Harry Robinson, Thomas King, James Welch, and Leslie Marmon Silko
Blanca Schorcht

On the Streets and in the State House
American Indian and Hispanic Women and Environmental Policymaking in New Mexico
Diane-Michele Prindeville

The Ecological Native

Indigenous Peoples' Movements and Eco-Governmentality in Colombia

Astrid Ulloa

Routledge
New York & London

Published in 2005 by
Routledge
Taylor & Francis Group
270 Madison Avenue
New York, NY 10016

Published in Great Britain by
Routledge
Taylor & Francis Group
2 Park Square
Milton Park, Abingdon
Oxon, OX14 4RN

Transferred to Digital Printing 2010

International Standard Book Number-10: 0-415-97288-4 (Hardcover)
International Standard Book Number-13: 978-0-41597-288-8 (Hardcover)
Library of Congress Card Number 2004024562

Library of Congress Cataloging-in-Publication Data

Ulloa, Astrid.
The ecological native : indigenous peoples' movements and eco-governmentality in Colombia / Astrid Ulloa.
p. cm. -- (Indigenous peoples and politics)
Includes bibliographical references and index.
ISBN 0-415-97288-4
1. Indians of South America--Colombia--Politics and government. 2. Indigenous peoples--Ecology--Colombia. 3. Indians of South America--Civil rights--Colombia. 4. Indians of South America--Colombia--Public opinion. 5. Public opinion--Colombia. 6. Colombia--Politics and government. 7. Colombia--Environmental conditions. I. Title. II. Series.
F2270.1.P63U55 2005
304.2'089'980861--dc22

2004024562

ISBN10: 0-415-97288-4 (hbk)
ISBN10: 0-415-88405-5 (pbk)

ISBN13: 978-0-415-97288-8 (hbk)
ISBN13: 978-0-415-88405-1 (pbk)

Taylor & Francis Group
is the Academic Division of T&F Informa plc.

Visit the Taylor & Francis Web site at
http://www.taylorandfrancis.com

and the Routledge Web site at
http://www.routledge-ny.com

For my mother and my daughter
Lilia and Naira

Contents

List of Tables

Acknowledgments

This text is only a part of a research process. During the process, I received support from many people and I am very grateful for their help. However, I apologize if I forget someone. I start by thanking professors from UCI (Bill Maurer, Mike Burton, Alison Brysk, Teresa Caldeira, Helen Ingram and Frank Cancian) for their comments, suggestions, and discussions during the whole process. I thank them for all their help and patience.

Support during my research was provided by the University of California, Irvine, the Inter-American Foundation, Instituto Colombiano de Antropología e Historia (ICANH), and Instituto Colombiano para el Desarrollo, la Ciencia y la Tecnología (Colciencias). Support during the writing process also came from a Regents' Dissertation Writing Fellowship, University of California, Irvine. I want to thank the Instituto Colombiano de Antropologia e Historia (their directors, María Victoria Uribe and Mauricio Pardo, and all my coworkers) for all their support in this process.

Sandra Durán, Luis Cayón and Juan Carlos Orrantia helped me in the collection of archival data. Claudia Campos contributed in the analysis of environmental implications of governmental and indigenous policies. Margarita Flórez helped me with the understanding of the international and national legal contexts. Andrés Barragán helped me with the preparation of the index. I am very thankful for all their help.

Walter Sondey (English Department, UCI) helped with my English style in the final version of this text, giving it a better textual touch. I thank him, for all his help and love.

Naira Bonilla (my daughter) was an important part of this process because she was always there for me. All the members of my family, Lilia Cubillos (my mother), Martha Ulloa, Zoraida Ulloa, Edgar Ulloa, Jorge Badel, Jim Brannies, Lynda Ulloa-Thompson and Andy Salazar helped and supported me throughout this process. They made my life easy and lovely. I am very thankful for having this wonderful family and I really appreciate their patience and help.

Finally, I thank the members of the indigenous organizations, Organización Indígena Gonawindua Tayrona-OGT and the Consejo Territorial de Cabildos-CTC of the Sierra Nevada de Santa Marta (SNSM), and their leaders Arregocés Conchacala, Cayetano Torres, Danilo Villafañe and Margarita Villafañe, who allowed me to participate in their own meetings and meetings with different institutions, and to discuss the indigenous peoples' political situations. I want to thank Milena Tafur, Eduardo Rico and Julio Barragán (advisors of the indigenous organizations) who always shared their perceptions about the situations of indigenous peoples of the SNSM. Álvaro Osorio and Margarita Bravo gave of their time, home and lives during my fieldwork in Santa Marta.

Chapter One
Introduction

POINTS OF DEPARTURE

The indigenous peoples of Colombia have become important and powerful interlocutors within national and international political arenas that allow them to rethink those political spheres and their predominant conceptions of nation, citizenship, democracy, development, and especially environment. Although indigenous peoples have been stereotyped during the last five hundreds years as savages, evil miscreants, or grown-up children, in the last few decades indigenous movements' political struggles have succeeded in transforming such representations. Many in Colombia and the international community now view indigenous peoples as *ecological natives* who protect the global environment and give us all hope in the face of the environmental crises brought about by western-style development. Consequently, representations of indigenous peoples in the developed world have changed from the "savage colonial subject" to the "political-ecological agent."

The main goals of this study are to analyze the construction of indigenous peoples' ecological identity; to reconstruct the historical emergence of the relationship between indigenous peoples and environmentalism; to assess critically the consequences that this relation creates for specific indigenous peoples' communities; and to determine how the cultural and environmental politics of indigenous peoples' movements have impacted national and transnational environmental politics by promoting different notions of nature and development. I am not looking for singular causes or trying to validate a theory, but rather I want to reconstruct the different networks, conditions of emergence, and implications (political, cultural, economic and social) of one specific event: the consolidation of the relationship of indigenous peoples and environmentalism.

In order to analyze critically and carefully the historical conditions and emergent environmental circumstances of indigenous peoples, I use Foucault's concept of eventalization. An 'event,' for Foucault, is a situation that expresses the beginning of a new and uncontested relationship, in this case that of indigenous peoples and environmentalism (as a global discourse which implies policies, knowledges, representations and practices related to environmental crises). Therefore, it is necessary to make explicit and problematic "the connections, encounters, supports, blockages, play of forces, strategies and so on which at a given moment establish what subsequently counts as being self-evident, universal and necessary" (Foucault 1991:77).

This study explores the relationship between indigenous peoples and environmentalism by looking for the multiple causalities, processes and interrelations that have helped to configure it. As Foucault writes, "Eventalization thus works by constructing around the singular event analyzed as a process, a 'polygon' or rather, a 'polyhedron' of intelligibility, the number of whose faces is not given in advance and can never properly be taken as finite" (1991:77).

There are no systematic studies that focus on the relationship between indigenous peoples and environmentalism in the concrete and historical manner presented here, although this relationship has been addressed before (Conklin and Graham 1995, Varese 1996b, Conklin 1997, 2002, Ramos 1998, 2002, Brosius 1999, 2000, Bengoa 2000, Brysk 2000, Ulloa 2001). My purpose is to contribute to the understanding of the links that currently exist between environmental identity construction and the political actions of the indigenous peoples' movements. I analyze the historical conditions of the emergence of the relationship between indigenous peoples and environmentalism as a process of coincidental development and consolidation that constructs a variety of ecological identities of both indigenous and nonindigenous origins that affect the manner in which indigenous peoples have access to political processes. This analysis necessarily involves the consideration of how transnational, national and local governments, nongovernmental organizations (NGOs), corporations, social movements and neoliberal models of development have affected the environmental discourses, representations and practices which increasingly influence the conditions under which indigenous people struggle for political, social, economic and cultural representation and power. This historical analysis of the representation of indigenous peoples as *ecological natives* thus allows me to address not only the local effects of environmentalism on specific indigenous groups, namely the Kogui people of Colombia, but also the effects that the *ecological native* in turn has had upon indigenous and nonindigenous peoples on the national and international scale.

A brief overview of scholarship on the emergence and activities of indigenous movements will help clarify the context of my argument. One of the most important arguments among scholars who study indigenous movements' political actions is how and why indigenous movements have recently gained political power within national and international arenas. Brysk (2000) argues that the indigenous movements' political actions have been successful because their identity and the international scope of their actions have achieved representation in international arenas such as the United Nations and human rights organizations. Yashar (1999) observes that under neoliberal policies indigenous peoples have successfully entered democratic political arenas to challenge those policies and traditional western conceptions of the linkage of democracy and development; while Alvarez, Dagnino and Escobar (1998) argue that identity construction and cultural politics allow indigenous peoples to propose new ways of doing politics. In addition, Varese (1996) considers this increasing political influence to be a result of indigenous peoples' resistance activities. Nonetheless, what these different analyses have in common is the recognition that indigenous peoples' movements have increasingly used political scenarios that were recently opened during the third wave[1] of democratization in Latin America to construct coalitions that have given them access to long-established national and international political arenas. In short, these scholars have shown how indigenous peoples, by gaining power in traditional political arenas, have transformed those arenas in the process and consequently changed their own identities and relationship within the structure of power. My goal is to build upon these findings and integrate them under the concept of the *ecological native* and to analyze the new political context in which indigenous peoples have gained multicultural and environmental rights.

In fact, at the international level indigenous peoples' rights have been recognized through legal international apparatuses such as the International Legal Organization's Convention No.169 (ILO-169). As Colchester states:

> International law now accepts that indigenous peoples enjoy collective rights: to ownership, control and management of their lands and territories; to exercise of their customary laws; to represent themselves through their own representative institutions. It is also recognized that laws, policies and 'development' should not be imposed on them without their prior and informed consent (Colchester 2002:2).

As this passage indicates, since the end of the 1970s, indigenous peoples' political actions and processes of constructing identities have responded to emergent ecological, environmental and conservation issues that have coincided with the process of the internationalization of environmental law. I argue that the emergence of the environmental crisis and environmental awareness has created a new political context and global environmental discourse based on expert scientific knowledge and new supranational environmental institutions. Among such institutions are the Convention on Biological Diversity (CBD) and the Global Environmental Facility (GEF), which is administered by the World Bank (WB). The discussion of the global environment in these circles proposes the management and regulation of the environment according to a view of biodiversity as "a world currency" (McAfee 1999). As the metaphor suggests, global environmental discourse has taken form in economic terms under the capitalist framework of the international markets.

In this context, indigenous peoples' movements seem to be a form of "empowerment," making indigenous peoples "free" to establish relations with international agencies and interests under "equal" conditions as social, political and economic agents, self-represented, autonomous, and with control over their territories and resources or "property." Private institutions may now interact directly with indigenous peoples, without state intervention, to negotiate the use of their "natural" resources. For example, in Ecuador a petroleum corporation has established direct ties with the Huaorani people, and a pharmaceutical corporation has done the same with the Awa people. Simultaneously, there are international policies such as those generated by the World Intellectual Property Organization's (WIPO) that are establishing policies for administering indigenous peoples' property rights. Consequently, the CBD's policies related to the recognition and protection of indigenous peoples' knowledges have been displaced by the WIPO scenarios. This new environmental context has given rise to new political events and relations that do not conform to previous analyses of indigenous movements' political actions.

I argue that indigenous peoples' political actions and the emergence of their "ecological" identities coincide with an internationalization of environmental laws and multicultural rights that have constructed indigenous peoples as subjects who can have full rights over their territories and resources only under the legal conditions and economic practices of the environmental marketplace. However, these situations and policies are still being formulated and can also be manipulated by indigenous peoples to propose changes and gain power within environmental and political discourses.

This possibility is already evident in the many instances in which indigenous peoples have proposed new forms of interaction between knowledge, nature and the management of natural resources that challenge globalization from the "top down" and that effectively constitute counter-globalizations or counter-governmentalities.

ECO-GOVERNMENTALITY AND ITS DISCONTENTS

In order to analyze the relationships between indigenous peoples and the environment, I use Foucault's concept of governmentality, which I develop as a concept of eco-governmentality. Then I explore how this eco-governmentality constitutes a response to global environmental discourses, multiculturalism, indigenous peoples' rights and global environmental governance.

The emergence of environmental awareness can be considered the birth of a new discursive formation (according to Foucault's concept of discourse) that produces a group of statements which provide a language for talking about—a way of representing knowledge about—the environment and ecological agents in a particular historical moment. Discourse in Foucault's perspective is concerned with representation as a source of the production of knowledge that in turn produces social practices and relationships of power. For Foucault, analyses of representations have to focus on the specific historical and social context of their production.

The global environmental discourse in question is the historical result of various texts, practices, conducts, policies and objects that share the same rules and premises for addressing the environment and, therefore, in Foucault's words, belong to the same discursive formation. Thus, the environment becomes a new space of knowledge that calls for special technical governance that promotes the beginning of a new specific governmentality (Foucault 1991a): an eco-governmentality.[2] For Foucault, governmentality refers to:

1. The ensemble formed by institutions, procedures, analyses and reflections, the calculations and tactics that allow the exercise of this very specific albeit complex form of power, which has as its target population, as its principal form of knowledge political economy, and as its essential technical means apparatuses of security.
2. The tendency which, over a long period and throughout the West, has steadily led towards the pre-eminence over all other forms (sovereignty, discipline, etc.) of this type of power which may be termed government, resulting, on the one hand, in the formation

of a whole series of specific governmental apparatuses, and, on the other, in the development of a whole complex of *savoirs*.

3. The process, or rather the result of the process, through which the state of justice of the Middle Ages, transformed into the administrative state during the fifteenth and sixteenth centuries, gradually becomes 'governmentalized' (Foucault 1991a: 102–103).

Watts (1993/94) summarizes the concept of governmentality as "all projects or practices intending to direct social actors to behave in a particular manner and towards specified ends in which political government is but one of the means of regulating or directing actions."

Following Foucault and Watts, I define eco-governmentality as all environmental policies, discourses, representations, knowledges and practices (local, national and transnational) that interact with the purpose of directing social actors (green bodies) to think and behave in particular manners towards specific environmental ends (sustainable development, access to genetic resources, conservation strategies and environmental security, among others). In this eco-governmentality, environmental organizations (governmental and nongovernmental), social actors (including indigenous peoples and their cultural and environmental politics), environmental activists and epistemic communities, among others, are agents in a process of regulating and directing social actions according to logics and discourses that contribute to the development of an emergent conception of global environmental governance. However, all these processes involve negotiations and conflicts as well as agreements.

Following Gupta (1998) and Luke (1999), I argue that this new eco-governmentality constructs international and national policies, discourses, representations and practices that introduce indigenous peoples to new production and consumption circuits. Gupta (1998) notes that "we may be witnessing the birth of a new regime of discipline in which governmentality is unhitched from the nation-state to be instituted anew on a global scale. In this project, global environmentalism comes together with other global accords and treaties, and the institutions through which these 'compacts' are monitored and enforced, to regulate the relationship between people and things on a global (not simply international) scale" (1998:321). In a similar way, Luke (1999) points out how "environments, therefore, emerge with biopower as an essential part of the constitution of modern 'man' who becomes the pretext for regulating life via politics" (1999:129).

Although environmentalism has generated various positions, tendencies and conceptions, it has tended to take the form of a global question in search of a unified global answer. Efforts to formulate such an answer generally take the form of proposals for control processes, which imply global actions that transcend local interests and conceptions, generating in this way a series of interdependencies between local and global contexts in which the global is the dominant force. This globalization process of environmental concern can be traced back to the 1960s. At that time, resolving human problems of pollution and extinction required not only new national solutions, new expert knowledge and new environmental practices, but also coordinated international action. Such a global perspective on environmental problems gave rise to universalizing strategies and responses. All humans (as ostensible equals without distinctions of class, gender, and ethnicity, among others) thus share a "common future" and the common task of resolving environmental problems. In response, multilateral organizations, international environmental NGOs, and transnational companies, just to name a few, have assumed the apparently "altruistic" task of saving the earth (Mother Earth) and promoting global processes to create more interdependence among countries as they seek ways to regulate and resolve environmental problems, such as population growth, food security and the loss of genetic resources. However, given that "first-world" countries lead this effort, the environmental crises have been mainly defined as problems caused by "third-world" countries and their unsound development and environmental policies. Thus, it seems that we are facing an eco-governmentality that determines daily environmental practices within an asymmetrical power relationship, and within this framework indigenous peoples are introduced to new processes of regulation, production and consumption in which their ecological practices and knowledges have begun to acquire representation and prominence around the world, but usually within a discursive context dominated by western perspectives, values and interests.

In this way, eco-governmentality is also related to multicultural discourses and policies. Hale (2002) notes that the representation of indigenous peoples in the context of "first-world" multiculturalism often subjects them to neoliberal assumptions and policies that offer the right of recognition at the price of conformity: "The state does not merely 'recognize' community, civil society, indigenous culture and the like, but actively, re-constitutes them in its own image, sheering them of radical excess, inciting them to do the work of subject-formation that otherwise would fall to the state itself" (496). One example of how multiculturalism can co-opt indigenous differences toward neoliberal ends can be seen in how ILO-169 eliminated some barriers that historically prevented indigenous peoples' attainment of full human rights. This convention recognized indigenous peoples' rights to self-determination and autonomy and initiated the requirement that they have political participation through indigenous organizations and representatives in the process of planning, discussing, and developing projects that affect their territories and lives. Practically, this right to participate has rarely resulted in indigenous control of the definition or outcome of the "development" to which they are subjected.

Correspondingly, since the 1970s, indigenous peoples have made efforts to represent, distinguish and promote their own environmental practices and development views at the national and international level. These indigenous peoples' environmental views began also to be related to global environmentalism, which prompted the construction of *ecological natives*. The recognition of indigenous rights and environmental rights are thus linked in a self-reinforcing relationship. Global environmentalism creates an opportunity for indigenous peoples to invoke the *ecological native* as a means to become authors and actors in their own discourses and claim that the contribution from their cultures to the nation and the world are the distinctive form of knowledge and respect that they have for the environment (Ulloa 2001). Consequently, two processes a "global" western environmentalism, and a recognition of the environmental and political rights of the indigenous, share many common features in terms of their hierarchical construction of power relationships, their perspectival limitations, and their contradictory relationship to one another regarding conceptions of property rights in natural resources (Flórez 2001).

I argue that we are engaged in the formulation of an eco-governmentality that assumes the need for global regulations and whose emergent discussions of biodiversity and sustainable development are necessary to the survival of the planet itself. Within the context of this emergent eco-governmentality, Colombia and indigenous peoples, in general, and the indigenous peoples of the Sierra Nevada de Santa Marta-SNSM, in particular, have taken a prominent position because their territories and "natural resources" encompass some of the hot spots of biodiversity that are focal points of this global environmental discourse.

The dominant tendency of eco-governmentality at present is to enforce the idea of the "ethnic group" as small nation conceived according to the ideas of territory and sovereignty of the nation-state model. This has allowed the definition of indigenous peoples as micro-nations that can deal with transnational corporations and transnational policies without nation-state control, but at the risk of assimilation to the neoliberal model. However, the construction of indigenous peoples' ecological identities may also be an opportunity to sponsor new ideas and alternatives to the neoliberal assumptions regarding territory, identity, autonomy, and nature.

The emergence of an eco-governmentality and its relationship to global environmental and multicultural policies imply new and contradictory situations for indigenous peoples of which three are pertinent in the present context: the relationship between national sovereignty and indigenous territories, self-determination and autonomy; the emergence of new conceptions of nature; and the linkage of indigenous peoples' rights and environmental rights.

One evident contradiction of many new international environmental policies is that they reinforce the idea of individual property rights in natural resources while also recognizing the collective autonomy of indigenous peoples. The recognition of biodiversity as a new commodity that can be valued, marketed and bought thus creates new political, economic and cultural situations for indigenous peoples regarding the management of their resources. These situations have increased indigenous peoples' presence in international arenas where the biodiversity of indigenous territories and the question of indigenous property rights carry problematic implications from a western perspective, particularly with respect to property rights among people who may not view nature as the possession of any one generation or as something that should or can be sold. This situation also raises issues related to the political autonomy of indigenous peoples and their sovereignty over their territories.

Biodiversity is geographically located largely in "third-world" nation-states, therefore "third-world" countries claim sovereignty over biodiversity and natural resources in indigenous territories to assure that transnational corporations have to deal with the nation-state rather than indigenous peoples. In the end, because biodiversity is seen as a natural resource, "third world" nation-states are reinforcing their idea of national territory and sovereignty in an effort to resist neoliberal transnational trade policies. Even though they often implement policies that break the model of nation-state unity and sovereignty by reinforcing decentralization and transnational participation at the local levels, as in the case of indigenous peoples, they also have to resist those policies to maintain their sovereignty over biodiversity.

This complex interplay of neoliberalism, international and national multiculturalism and environmentalism forms the dynamic context in which indigenous peoples' rights and global environmental laws based on property rights designed to protect biodiversity come into conflict with the idea of national sovereignty over natural resources. In Colombia, for example, national laws recognize the idea of multiculturalism and indigenous peoples' rights to decide the management of natural resources in their territories, which means that those peoples could negotiate directly with transnational corporations and erase the state's power. Indigenous peoples are correspondingly empowered to negotiate as nations with transnational corporations regarding their natural resources, which contradicts neoliberal policies of the reduction of states' power by creating (recognizing) multiple micro-nations with power on a micro scale.

Another contradiction arises from differing definitions of nature itself. The idea of genetic resources has taken two forms in environmental discourses: as a "natural thing" and as a "constructed thing." These conceptions are different but have similar implications for indigenous peoples. If genetic resources are "natural things," they belong to humanity; therefore indigenous peoples have to share these natural resources or manage and preserve them for others. If genetic resources are socially constructed, they belong to indigenous peoples; therefore they can market their resources as a commodity. The recognition of indigenous peoples' rights may thus permit a more western form of environmental ownership and industrial management of resources, or it may also imply reinforcing indigenous tradition as an element of resource management, thus promoting the making of inventories and collections that resemble the colonial processes evident in museums and market-oriented inventories. In this latter case, indigenous people become the managers or caretakers of their own cultures and territories under marketing or preservationist assumptions originating in the West.

The final contradiction between indigenous peoples and environmentalism is related to rights of recognition. Indigenous peoples' rights and environmental rights have been ratified in international and national contexts in a parallel way so that international environmental laws and indigenous peoples' laws are now totally interrelated on the basis of western legal conceptions. However, the practical development and the implementation of these rights have been slow and sometimes ignored. Juridical analyses show that the basic notions that sustain the rights of indigenous peoples are not applied fully because in many nation-states the indigenous are not completely recognized as sovereign *peoples*. Further complicating the recognition of indigenous property rights are western environmentalists' conceptions of sustainable development that often do not correspond to indigenous conceptions of environmental management so that the indigenous peoples themselves have little interest in asserting their rights under those terms.

According to Gupta (1998) and McAfee (1999), the recognition of indigenous sovereignty is often linked to the economic potential of indigenous resources. In this sense, indigenous peoples' knowledges and territories are not recognized for their biological or cultural value, but rather for their market value. Moreover, it seems that when nature has become a global currency (Gupta 1998, McAfee 1999, Sachs 1999), the practices and knowledges of indigenous peoples are recognized only when they acquire a value in this new "free" eco-market, as in the case of providing bio-prospecting services or cultural spectacles and eco-commodities for eco-tourists.

The contradictions and inequalities of this power dynamic can be seen in the way indigenous peoples have to conform their traditional practices to national standards of ecological activity that reproduce international patterns of sustainable development based on notions of the "free" market (McAfee 1999). In fact, many sustainable development projects that have been introduced in "third-world" countries impose a global perspective on the management of natural resources without giving consideration to local indigenous practices and strategies.

These contradictions and emergent relationships suggest various general questions: Who will deal with these new entities (indigenous peoples' territories and autonomous political entities)? In what arenas will the interrelation of indigenous peoples and multinational corporations occur? How can multilateral institutions be formed that include indigenous peoples on more equitable terms?

However, this general overview also suggests other, more specific questions that address the combined influence of environmentalism and neoliberalism on indigenous peoples: Are indigenous peoples gaining or losing ground through inclusion in global ecogovernmentality? What will happen if they negotiate all their natural resources and even territories for international eco-tourism or eco-products? What will happen to their "unique identity" after it has been consumed by international markets? What are the strategies to deal with new contexts in which all natural resources can be bought or sold? What will indigenous power be based on? What are their options? If they participate in the green market (ecological products), how long will the ecological era last? (In fact, some have said that the ecological era is ending.) Finally, are they gaining an opportunity to propose counter-globalizations, counter-governmentalities and alternative modernities, or are they simply subjecting themselves to the terms of a hegemonic western eco-governmentality?

OUTLINE OF THE TEXT

In this context, understanding indigenous peoples' movements entails analyzing the indigenous peoples' historical construction of ecological identities and their new relationships with national and transnational environmental movements and global environmental laws. It also necessitates an analysis of how, why, and in what ways indigenous peoples' movements are involved in neoliberal policies and global environmental law.

To develop my arguments, I analyze the emergence of indigenous peoples' movements in Latin America and Colombia not only as a result of indigenous political power but also as responses to transformations generated since the 1970s by democratization and globalization processes; and I show how these are linked to the growth and the spread of communications technologies which connect local-global processes and transform the temporal and spatial situations of local social movements by articulating them to transnational scenarios. At the same time, indigenous peoples' circumstances are linked to the transformation of the state by decentralization and the implementation of neoliberal policies (including the privatization of state institutions, the abolition of subsidies, and the opening of the Colombian market to the international market), and specifically neoliberal multiculturalisms, in Hale's (2002) terms.

Throughout this text, I explore the history of indigenous peoples' movements as well as environmental movements. I examine the construction and confluence of these two movements and how their relationships give rise to new ideas of identity, representations, territory, autonomy and natural resources that have engaged indigenous peoples and prompted their actions within a specific framework of ideas (eco-governmentality) that relate them to nonindigenous practices, discourses, representations, political participation, territory, and concepts of nature and property rights, among others.

This text is based on ethnographic information and it is focused on the historical process of the emergence of indigenous peoples' movements in Latin America, in general, and those of the indigenous peoples of the SNSM, in particular. It demonstrates the processes of the construction of indigenous peoples' environmental identities as an interplay of local, national and transnational dynamics among indigenous peoples, political entities, NGOs and environmental movements and their discourses in the context of global environmental policies; and it also demonstrates the specific political effects of these processes and discourses upon indigenous peoples' autonomy, territories, resources, knowledges, identities, and representations.

For this analysis, it is important to understand the specific processes by which the SNSM indigenous peoples' cultural and environmental politics and their proposals have reached national and transnational political spaces and successfully created dialogues with both governmental and nongovernmental organizations. It is this interaction among cultural dynamics, organizational processes and cultural and environmental politics that has generated arenas for the representation, negotiation and consolidation of indigenous people's interests and values. At the same time, indigenous peoples are also being consolidated within these arenas as *ecological natives* within the context of a global eco-politics and eco-governmentality. As I demonstrate, this struggle for representation and power creates contradictions or is itself the effect of the contradictory processes and contexts in which it takes place.

Chapter 2, "The Emergence and Continuity of Indigenous Peoples' Movements: Latin America and Colombia," draws on social movements theories, specifically Alvarez, Dagnino and Escobar's (1998) concept of cultural politics, and Brysk's (2000) analysis of indigenous peoples' movements' identity and internationalization. I do so to analyze the emergence and continuity of indigenous peoples' movements' environmental actions and identities and how they affect the dominant political order as a result of their cultural politics, which often seek to change or disrupt that order by constructing transnational networks. I therefore address the emergence and development of indigenous peoples' movements in Latin America and Colombia to analyze not only their own political systems and their opportunities and constraints, but also to analyze the political actions and trajectories of these movements in terms of their interrelation with institutional political systems at the national and international levels. This approach allows me to identify how the interrelations of national and transnational networks have an impact within specific national and transnational political arenas and vice versa. In short, I concentrate on the influences these movements and their organizational dynamics have upon one another. I also analyze how multiculturalism and the recognition of difference may be viewed as elements in the process of constructing nationalism (Colchester 2002, Wade 1997, Bhabha 1994) and how the inclusion of minorities may only be a new way of controlling these minorities who could not otherwise be integrated into the nation-state or the neoliberal policies (Ortíz and Hernández 1994, Motzafi-Haller 1995, Gros 1998).

In chapter 3, "Indigenous Peoples' Movements of the Sierra Nevada de Santa Marta, Colombia," I focus on the emergence and consolidation of the indigenous peoples' movements, as well as the organizational processes and political struggles of indigenous people of the SNSM and their relation to broader indigenous political actions within the Latin American and Colombian contexts. Indigenous peoples of the SNSM have experienced a particular organizational process through the formation of their own indigenous organizations, specifically the Gonawindúa Tayrona Organization (Organización Gonawindúa Tayrona-OGT)[3] and the Cabildo Territorial Council (Consejo Territorial de Cabildos-CTC[4]). In this chapter, I show how the CTC's proposals promote an indigenous environmentalism and territorial knowledge that permeate the proposals of a series of local, regional and transnational actors (mayors' offices, corporations, NGOs, state institutions and multilateral institutions, among others) that are incorporating indigenous cultural meanings and practices into environmental discourses. These interactions between the CTC and the state institutions have established agreements and differences and built new strategies that affect territorial, cultural and environmental issues and thereby reposition indigenous peoples of the SNSM as agents able to propose new and different territorial, cultural, and environmental relationships. This chapter presents the indigenous peoples' cultural politics and gives a basic ethnographic setting in order to situate analyses in the following chapters.

Chapter 4, "Thinking Green: Global Eco-governmentality and its Effects in Colombia and the Sierra Nevada de Santa Marta," analyzes the emergence and development of global environmental movements. This chapter is a journey though the emergence and development of global environmental awareness. Environmental discourses are explored with a particular emphasis on two outstanding tendencies: the anthropocentric and the biocentric. Then, I address how these tendencies enter environmental discourse and become part of the global environmental strategy (eco-governmentality) that has influenced Colombian environmental policies (Foucault 1991, Gupta 1998, Luke 1999, Escobar 1999). Finally, the chapter addresses the development of environmental discourses and movements in Colombia and, in particular, in the SNSM with regard to their interrelation with global environmental policies.

In chapter 5, "Ecological Identities of Indigenous Peoples: Historical Process of Construction," I explore the historical conditions of the emergence of the relationship between indigenous peoples and environmental movements and the construction of ecological identities (Hall 1990, 1997, Scott 1995, Conklin and Graham 1995, Varese 1996b, Wade 1997, Castells 1997, Comaroff and Comaroff 1997, Conklin 1997, 2002, Ramos 1998, Brosius 1999, 2000, Bengoa 2000, Brysk 2000, Ulloa 2001). Drawing on identity construction theories, I analyze indigenous ecological identities as the result of intricate and dialectical links between indigenous peoples' movements' cultural politics and global environmental policies. Indigenous traditions (territoriality, historical memory and a quotidian sharing) and new political strategies (new forms of organizations and a fluid identity) are examined as mechanisms that indigenous peoples use to enact the historical continuity of their traditions and identities and respond to changes that, in turn, produce political changes at a global level. Thus, I analyze ecological identities as the products of different contradictory discourses generated by various agencies situated at different points on the spectrum of power, which include: indigenous peoples' cultural politics and traditions, transnational entities, expert knowledge, environmental NGOs (ENGOs), indigenous organizations, the state, and neoliberal policies, among others.

In this chapter, I discuss how the new articulation of indigenous peoples and the environment not only constructs new identities but also causes new concerns related to territory and autonomy that differ from earlier indigenous struggles. Emergent indigenous claims to autonomy and territorial rights arise not only from within the dominant forms of political organization of the nation-state that enjoys autonomy within its own borders, but also from the transformation or reduction of the nation-state's autonomy and sovereignty in the context of a global and transnational eco-governmentality. Indigenous peoples' demands thus imply a new relation to natural resources and access to genetic resources due to their desire for sovereignty over their own territories. Therefore, I examine the environmental politics of indigenous peoples' movements in a manner that articulates territory and environment within and among transnational, national and local power dynamics.

In chapter 6, "Environmental Images and Representations: Implications for Indigenous Peoples," I show how the images of *ecological natives* have become an important political strategy for indigenous and environmental movements. However, these identities are in a constant interrelation with stereotypes of otherness that has impacted national and international representations of indigenous peoples (Said 1978, 1997, Nochlin 1989, Torgovnick 1990, Lutz and Collins 1993, Webb 1995, Hall 1997, Maxwell 1999). In this chapter I explore different representations, images and associations that historically have been related to indigenous peoples, such as Mother Earth or the "noble savage." I argue that these images have oscillated within the nature/culture dichotomy of the western paradigm and resemble colonial processes. I also demonstrate that transnational environmental indigenous peoples' movements have often helped create such essentialized images of indigenous peoples.

Chapter 7, "The Power of Ecological Identity: Alternative Ways of Thinking and Acting within a Globalocality," shows how indigenous peoples' struggles related to the environment and their ecological identities have been transformed into effective means to appeal to transnational entities and establish bonds, coalitions and networks with them (from financial help to political and conceptual support) that give the indigenous peoples more political power within nation-states. Indigenous peoples thus transcend not only territorial borders (as do migrants and diasporas), but also cultural borders. By crossing the representational and discursive boundaries of other cultures, indigenous peoples establish new political relations and different memberships and identities with international regimes, international social movements, and other indigenous cultures, to name just a few examples. Moreover, indigenous practices demonstrate that transcultural and transnational encounters have been long-term components of their cultural dynamics. These multiple and dynamic identities and loyalties (conservationist, NGO, and indigenous nations, among others) situate indigenous peoples' movements within a new dimension of citizenship within the nation-state. Consequently, indigenous peoples' actions and identities and their interrelation with different social actors have socioeconomic and political implications for other actors according to their specific social locations. I focus on how the cultural and environmental politics of indigenous peoples are affecting these actors, including the state, multilateral institutions, transnational corporations, environmental non-governmental organizations, social movements, local actors, and researchers.

Chapter 8, "Indigenous Peoples within Eco-Governmentality," shows the contradictions that complicate the relationship between indigenous peoples and environmentalism. I examine how indigenous peoples' actions are the result of their empowerment as actors in social movements and how their political actions have been recognized by states and transnational institutions. Indigenous peoples' struggles to define and control their identities have given rise to important strategies that have appealed to transnational actors by establishing bonds, coalitions and networks that have given indigenous peoples more political power within nation-states. However, I argue that within the new environmental regimen or eco-governmentality, neoliberal policies are reaffirmed to the extent that indigenous peoples need to be free and autonomous agents in western terms in order to negotiate their territories, resources, and knowledge that now are valuable in environmental markets.

CONCEPTUAL AND METHODOLOGICAL APPROACH

My basic methodology was to interact with indigenous peoples through local participation in research and with local notions and processes of communication in order to establish a real possibility of reciprocity among the "research subjects" and the researcher. In this sense, I believe that indigenous peoples are researchers. Thus, the practices of investigation for this text became a collective construction in which indigenous peoples and I, as a "researcher," established common interests, types of exchange and methodologies to collect the data. For me, participant observation has a different meaning from the traditional Malinowskian framework. I understand it is as a collective enterprise that follows common interests. This project was therefore based on local participation in which opinions, feelings, beliefs, as well as more global ecological and political factors, were all taken into account.[5]

In order to construct a common enterprise with indigenous peoples, I followed the methods of the so-called "Interactive Participation" that were developed by Ulloa, Rubio and Campos (1996a, 1996b, 2000) for addressing environmental issues. These methods allow the inclusion of all the different social actors involved in particular research projects in the process of defining the terms of participation, meetings among actors and the means of socializing the information that these actors may generate. The goal is to respect indigenous peoples' rights to self-representation and sovereignty over their territories and to allow them to decide when, how, and with whom they want to work. This required that I discuss my purposes and plans with them.

The right to self-representation has different implications for indigenous peoples' relations to national policies, the processes of research regarding natural resources, and projects that affect the lives or territories of indigenous peoples. In this sense, research topics and field research are collective decisions. This perspective prevented me from applying preestablished methods; rather, it opened an arena for constructing different methods according to the ad hoc development of the project. Consequently, I participated in their meetings (workshops, meetings with national and local governmental authorities and nongovernmental organizations), and we collectively arrived at methodologies that suited our mutual projects.

It is important to highlight that in Colombia most indigenous peoples forbid any research in their territories, and they specifically forbid biological research. Some indigenous peoples' organizations have special guidelines for doing research. In these general proposals they state the necessity of recognizing the collective intellectual property rights of indigenous peoples. For this research, I assumed the principles that the OGT and the CTC have for anthropological research. I worked almost one year with the CTC in order to construct a way to perform research together. After that, I did fieldwork for 15 months, going back and forth between Bogotá and Santa Marta.

As a second methodological approach, I focused on the role of different actors at the local and global levels.[6] I also analyzed all actors and their political actions and the interrelations among them that shape indigenous peoples' conceptions and practices. Moreover, I focused on ways that these actors managed collaborations and agreements as well as conflicts and asymmetrical power relations. Understanding these bonds and links allowed me to define indigenous peoples' concerns and challenges at the local, national and global levels.

Under this approach, it was necessary to rethink concepts of space and place in ethnographic research, especially as related to cultural formations in this case: indigenous peoples' movements and their relations to environmental discourses. To trace the national and transnational relations of the indigenous peoples' movements required a fieldwork that addressed the circulation of people, things, ideas, representations and histories through different spaces, contexts, languages, cultures, and everyday practices. Therefore, I used multi-sided ethnographic techniques that permitted a new conception of spatiality and the linkages among sites (Marcus 1995). My analysis of indigenous peoples' political and ecological practices derives from field research in which I recorded, observed and participated in the indigenous movements' activities at the local level and from participation and research at the regional, national and transnational levels among nonindigenous actors (governmental institutions, NGOs, grassroots, and multilateral institutions, among others). Finally, I gathered information among different actors through archival research, newspapers, interviews, life histories, and participant-observation.

This conceptual and methodological approach was a result of the application of the abovementioned theoretical and methodological perspectives and direct engagement with indigenous peoples. It helped to open a space of discussion about the role of anthropologists within these societies, and it allowed me to develop new ways of understanding indigenous peoples' concerns and challenges.

The academic goal of my work is to contribute to the critique and elucidation of the power/knowledge relationships that indigenous peoples experience when they engage in eco-governmentality. My ultimate goal, however, takes shape as the hope that this text may provide indigenous peoples yet another opportunity to express their thoughts, dreams and struggles.

Chapter Two

The Emergence and Continuity of Indigenous Peoples' Movements: Latin America and Colombia

INTRODUCTION

Indigenous peoples collective resistance activities and political interventions in Colombia and Latin America began in colonial times. However, it was not until the 1970s that indigenous peoples succeeded in consolidating their presence within national and international arenas as formally recognized political agents. As such, they have gained the power they need to change laws and political discourses related to rights, economic planning, development, the environment and democratization formerly denied to indigenous peoples in Latin America and Colombia as a matter of customary tradition or formal political exclusion.

The indigenous peoples are now active participants in contemporary negotiations and discussions related to their territories, natural resources and knowledges in national and transnational political arenas. They now participate in the electoral arena, generating new ideas, running for office and even influencing the outcomes of presidential elections in Latin America. In addition, indigenous peoples have sparked action among nonindigenous sectors such as students, workers, and environmentalists. In this way, they are constructing alliances, real or symbolic, that have changed the perception and reality of the role and the power of indigenous peoples in the political arena.

Social movements' theories are generally useful in the analysis of indigenous peoples' collective actions, especially those that have focused on indigenous actions that have occurred since the 1970s. I will use a theoretical perspective on social movements that defines them as a result of collective actions that aim either "to breakdown the social system" or form "new interests or new forms of solidarity and collective identity" (Melucci 1980: 212). Indigenous social movements have often focused on cultural identity as a political end in itself, which has resulted in new forms of solidarity and new interests. This focus on cultural identity has produced notable changes in personal and social relations throughout Latin American nations. Moreover, social movements' actions have promoted a democratization of the institutions of civil society that has given rise to new ways of doing politics. Consequently, the collective processes of constructing identities have made social movements serious forces for change within national political arenas.

Social movements' political actions are related not only to broader institutional systems, but also to multiple local interactions introduced through various media and embedded in the daily-life practices of individuals. The notion of political action consequently expands to include the capillary effects of media processes by which meanings spread through social movements' networks to affect individual consciousness. From this perspective, the effects of social movements in Latin America can be seen both in their formal political consolidation within the democratization process and their influence upon informal social and cultural processes. Accordingly, the efficacy of social movements in Latin America can be assessed not only by their impact on the political process of democratization but also by their role in the strengthening of civil society and cultural activities that promote individual political participation (Alvarez, Dagnino and Escobar 1998, Dagnino 1998, Warren 1998, 1998a, Escobar and Alvarez 1992, Nash 2001, Pardo 1997, Brysk 2000, Edelman 2001, Ramos 1998, Archila and Pardo 2001).

From this perspective, the activities of indigenous movements' can be analyzed in a different way. In fact, the question is not only why they arise or how they act, but also how the cultural politics of indigenous movements and their construction of identity have changed and resignified politics by opening new democratic arenas and changing the ways of doing politics. An understanding of indigenous strategies for constructing collective identities through social movements is therefore necessary to the analysis of how they have changed national and, ultimately, transnational political arenas.

Studying social movements, specifically the indigenous movements and their emergence, development, and eventual decline, involves close analysis of the interrelation of political opportunities, mobilizing structures, and cultural factors (framing processes) at the local, national and transnational levels. It also implies analyzing the way that individual actors construct their identities and talk about their daily lives in addition to analyzing the general social dynamics of their practices in relation to broader historical processes of power/knowledge. Moreover, it implies that social movements are key motivators of the social process of identity formation that reinterprets norms, constructs new meanings, reshapes the public discourse, challenges boundaries between public/private spaces and cultural/political fields, and empowers social movements to create new ways of doing politics in which cultural differences and identities are powerful factors. Furthermore, it involves the analysis of how various individuals and indigenous groups have effectively consolidated themselves to form political blocs capable of collective action, as opposed to an analysis that focuses on the negatives: those instances in which the indigenous are dominated as disenfranchised groups or victims (Cohen 1985, Touraine 1985, Escobar 1992, Wilmer 1993, McAdams, McCarthy and Zald 1996, Tarrow 1998, Brysk 2000).

In Latin America, social movements have been analyzed as political actors that have helped to change notions of identity and law by challenging traditional authoritarian orders. Among these social movements, indigenous movements may be counted among the agents that have done the most to reconfigure established national conceptions of identity, development, democracy and nature. Influential studies[1] have presented the general impact and scope of indigenous movements' political actions within national and transnational arenas in Latin America. However, there are none that focus on Colombia and consider how the environmental politics of indigenous peoples' social movements there have had an impact on national and transnational environmental discourse and policy and vice versa.

Within this context, and drawing on Alvarez, Dagnino and Escobar's (1998) concept of cultural politics and Brysk's (2000) analysis of the indigenous peoples' movements identity and internationalization, the following pages will focus on Colombia as a prime example of how the indigenous movements' identity politics can affect the dominant national political culture by constructing national and transnational networks that change or disrupt the terms of its power.

In this chapter, I focus on the emergence and development of indigenous peoples' movements in Latin America and Colombia in order to analyze not only their political systems as such, but also to document the practical history of their political actions. This historical approach allows me to identify and analyze the impact of indigenous movements on specific processes and activities within transnational political arenas. It is necessary to clarify that in this chapter I talk about indigenous movements and their actions in general, and when I use the term "indigenous movements" I mean grassroots organizations and their actions as well as leaders' actions or individual actions. In short, the aim of this chapter is to present pertinent theoretical approaches to social movements from the social sciences in order to form a context for recounting the historical emergence of indigenous peoples' movements in Latin America and Colombia and to make some remarks about the positioning of indigenous peoples' movements in the western political imagination in general.

THE EMERGENCE OF SOCIAL MOVEMENTS THEORIES

New social movements were "discovered" by the new or contemporary social-movements theory that emerged around the mid-1970s as part of an effort to understand collective actions that occurred during the 1960s and 1970s in the United States and Europe,[2] such as the civil rights and student movements (1960s), the feminist and ecological movements (1970s), and the peace and local autonomy movements (early 1980s). These social movements were responses to quality-of-life issues and cultural and ethical problems, and they were led by actors whose social situations were before unknown to the dominant actors in state contexts. As Epstein notes, the goals of new social movements included "cultural revolution, within and without: that is, creating a movement that embodies the values of a new society and brings about fundamental changes in social values, in the way that people think about social relations" (Epstein 1990:36).

These movements are characterized by their inclusion of new social actors in political processes (students, minorities and women, among others), radical forms of action (occupations, interruptions, direct actions, civil-disobedience and violent confrontations), antimodern values, decentralized organization (different from party systems and representative associations), rejection of institutional politics, and claims to greater democratization of social structures within civil society. These features, according to the new theory, distinguish the new movements' purposes from older conceptions of social movements' goals: as means to address conflicts primarily related to economic and labor problems. Theoretical distinctions began to be drawn because these movements were not related (at least on the surface) to socioeconomic class struggles, and because classical sociological theories (Functionalism and Marxism) proved inadequate to understand such massive, novel social actions (Cohen 1985, Offe 1985, Touraine 1985, Klandermans and Tarrow 1988, Scott 1990). To summarize, the new social movements can be defined as a result of collective actions that try either "to breakdown the social system" or to form "new interests or new forms of solidarity and collective identity" (Melucci 1980:212). As Touraine notes, "The idea of social movements introduces a different approach because it tries to evaluate the capacity of various categories to transform themselves into actors of their own situation and of its transformation" (Touraine 1985:783).

Within social-movements theory, generally speaking, there have been two main paradigms or approaches to the analysis of collective actions: in the United States, the "resource mobilization" approach; and in Western Europe, the "identity-oriented" approach. In the United States, the "resource mobilization" approach focuses on "the group and the individual level, looking systematically at the groups that organize mass protest, at their forms of action, and the motivations of individuals who joined them" (Klandermans and Tarrow 1988:3). This analytical paradigm addresses collective action in relation to the availability of resources and political opportunities. In other words, these researchers focus more on "how" these movements act. Such studies emphasize three important elements: the cost-benefit analysis of participation, organization, and expectations of success (Klandermans and Tarrow 1988, McAdams, McCarthy and Zald 1996, McCarthy, Tarrow, and Tilly 1997).

In Western Europe, the "identity-oriented" or "new social-movements theory" focuses on cultural meaning and identity formation, especially "larger structural issues—the structural causes of social movements, their ideologies, their relation to the culture of advanced capitalist society" (Klandermans and Tarrow 1988:2). This European approach focuses more on "why" these movements come about (Melucci 1980, 1985, 1993, Touraine 1985, Klandermans and Tarrow 1988). In this paradigm, collective action is analyzed in relation to values (antimodernism), action forms (unconventional nonhierarchical organizations), constituency (groups of marginal people), new aspirations (postmaterial values), and satisfaction of endangered needs (increase of social strain). Studies in this theoretical framework have analyzed these values, action forms and constituencies to identify the structural preconditions of their emergence and power, as well as their relation to postindustrial society. According to the terms of Touraine's analysis these transformations in the organization of production and cultural models produce new dimensions of power relationships that generate the new kinds of subjectivities and identities that characterize postindustrial society. Some scholars (Epstein 1990, Plotke 1990, Escobar 1992) argue that the "New Social Movements" paradigm is the most relevant for understanding collective actions because of its emphasis on culture and ideology as important terrains and motives for political struggle. Other scholars argue that, though the European and American approaches have differences, they are complementary. For present purposes, these two approaches will be integrated in order to analyze how and why indigenous social movements have arisen, with a particular emphasis on issues concerning structural changes and their effects on organization, politics and resources involved in the mobilization processes that have brought them to power.

Both approaches propose that "social movements involve contestation between organized groups with autonomous associations and sophisticated forms of communication (networks, publics)" (Cohen 1985:673). Moreover, both approaches point out that in order to understand social movements, it is indispensable to focus on *civil society* because it is on that terrain that social actors mobilize and seek identity, autonomy, and recognition through their struggle to create new kinds of social and political relations. In this way, the motives and actions of individual subjects become relevant to understanding large-scale political dynamics.

Following Cohen (1985), one of the salient features of social movements is not that "they engage in expressive action or assert their identities, but that they involve actors who have become aware of their capacity to create identities and of power relations involved in their social constructions" (Cohen 1985:694). Social movements are therefore related to ideological discourses and social networks insofar as individual and collective action is culturally inscribed and communicated. These movements are characterized not only by their collective capacity to challenge, in a sustained and disruptive way, other groups, elites, authorities, powerful opponents, but also by their capacity to modify discourses and reinscribe cultural codes (Melucci 1980, Offe 1985, Cohen 1985, Kauffman 1990, Scott 1990, Escobar and Alvarez 1992, Wilmer 1993, McAdams, McCarthy and Zald 1996, Tarrow 1998).

McAdams, McCarthy and Zald (1996) point out how the various theoretical approaches to the analysis of the emergence of social movements have three emphases in common: political opportunities, mobilizing structures and framing processes. The first of these elements emphasizes how social movements' politics are related to the opportunities and constraints of a broader institutional political system: "Social movements and revolutions are shaped by the broader set of political constraints and opportunities unique to the national context in which they are embedded" (McAdams, McCarthy and Zald 1996:3). And so social movements occur when "changing political opportunities and constraints create incentive for social actors who lack resources of their own" (Tarrow 1998:2). This implies that the state does not simply rule civil society, but that it is in a dynamic and responsive relationship to social actors whose "political" opportunities, under circumstances of political constraint or repression, often arise outside of institutional politics and officially recognized political arenas. In this political interrelation, social movements may express their aims through a varied repertoire of contention that promotes mobilization, protests, obstructions, petitions, assemblies, strikes, and marches. (Tarrow 1998).

The second common element addresses the forms of organization (formal and informal) that these groups use, particularly the forms of organization available to them within and outside the institutionalized political system and already-established social networks. Among these forms of organization are intellectual circles, grassroots organizations, churches, informal friendship networks, and cultural practices. Some scholars argue that knowing these dynamics of organization allows closer analysis of the relationships between the type of organization and kind of movement as well as the organizational infrastructures within a specific country and the effects of the state structures and organizational cultures on the social movement (McAdams, McCarthy and Zald 1996). Tarrow (1998) notes that it is ultimately the practical activities of individuals or the "life within groups that transforms the potential for action into social movements" (22).

The third common element for the analysis of emerging social movements is the attention paid to framing processes: the examination of the shared feelings and meanings that motivate individuals to seek political opportunities and mobilizing structures and to form a common identity that catalyzes collective action. In other words, "Mediating between opportunity, organization, and action are the shared meanings and definitions that people bring to the situation" (McAdams, McCarthy and Zald 1996:5). This is the cultural framing process that allows social actors to legitimize and motivate collective action and which Snow defines as "the conscious strategic efforts by groups of people to fashion shared understanding of the world and of themselves that legitimate and motivate collective action" (McAdams, McCarthy and Zald 1996:6). Tarrow (1998) states that these collective-action frames allow the identification of "us" and "them," which creates the dialectical relationship that can promote a mutual transformation of identities.

Social movements have been analyzed primarily within national contexts. However, other scholars (Brysk 1993, 1994, 1996, 2000, Wilmer 1993, Santos 1998, Tarrow 1998, Alvarez, Dagnino and Escobar 1998) have addressed the impacts that these movements have had on transnational political arenas and transnational political factors (opportunities, mobilizing structures and framing processes) that in turn affect the configuration, emergence, development, and effects of social movements. These transnational mobilizations allow cooperation across borders, construction of transnational identities and the use of transnational mechanisms, such as international law, to transform their political situations. At the same time, as Brysk (1996) points out, international coalitions help to transform national policies and politics.

Correspondingly, national and transnational relations influence the internal development of social movements (McAdams, McCarthy and Zald 1996, Tarrow 1998). The political institutions that form the arena in which these movements operate change the political opportunities and reshape the structure of the social movements that enter into them. Social movements thus change their initial mobilizing structures (for example, institutional and informal networks) to give birth to new networks of relations or to more permanent organizational structures (formal organizations) as a result of at least three factors: the disruptive tactics of external political forces and processes, intragroup divisions resulting from the radical flanks effect, and the dynamics of formal political goal-setting (nondisplacing and single-issue). Moreover, social movements mutate as they create allies and consensus with other traditional and nontraditional political organizations and parties in their efforts to accomplish their goals. Finally, framing processes (cultural dimensions) become more formally organized as an explicit strategy to reaffirm the identities of the movements and to establish them as well-recognized and serious forces for social change within the political arena, thus promoting confrontational processes between long-established interests within the state (Melucci 1985, 1993, McAdams, McCarthy and Zald 1996).

Although cultural dimensions have been pointed out as essential factors in the emergence and development of social movements (Melluci 1980, Touraine 1985, Epstein 1990), there has been less systematic work on them (McAdams, McCarthy and Zald 1996, Escobar 1992). For this reason, Escobar (1992) claims there is a need for "the reconstitution of meanings at all levels, from everyday life to national development" in order to grasp the cultural politics of these social movements. As I demonstrate in the next section, cultural dimensions are indeed a central interest within Latin American theories about social movements.

SOCIAL MOVEMENTS AND SOCIAL MOVEMENTS THEORIES IN LATIN AMERICA

In Latin America since the 1970s, a multiplicity of new actors has appeared on the public stage promoting new forms of identification in new political arenas. Urban inhabitants, peasants, workers, indigenous peoples, women, mothers, ecologists, ethnic minorities, and many others, have risen up against their problems to demand a clear change within the social, economic and political spheres of their lives.[3] Some scholars argue that the beginnings of these social protests resulted from economic and political crises: the crisis of development (as a cultural and economic discourse) that did not "improve" the social condition of most of the population of Latin American countries, and the crisis of the loosening of constraints on political representation and participation that forced established political parties and interests to the open authoritarian regimes to new political actors and their interests (Escobar and Alvarez 1992, 1992a, Calderón, Piscitelli and Reyna 1992, Wade 1997). Other scholars argue that these movements are reactions to the capitalist transformation and modernization of social relations, particularly globalization and neoliberalism. However, in Latin America, such modernization developed under specific historical and social processes, and thus these social movements differ from American and European movements (Calderón, Piscitelli and Reyna 1992).

These collective actions, which demanded political recognition of the individual citizen's rights and promoted democratic values, arose from constituencies whose identities were not based on class but whose claims were nevertheless directed toward increasing popular participation in the democratic processes that regulate economic and social activity. Thus, the old left theory that considered society as an unchangeable structure that only class struggles could transform by taking control of the state could not explain them. Moreover, some scholars (Escobar 1992, Calderón, Piscitelli and Reyna 1992, Wade 1997, Pardo 1997) believed that Latin American analytical approaches articulated under dependency and modernization theories were also inadequate to explain these movements because of those theories' ideological commitments and limitations, and particularly because they focused on structural relations and linear causality rather than on multiple processes of transformation. These paradigms also identified external forces as the causes of processes and activities among peripheral groups attempting to gain access to centers of power, which largely precluded analysis of local initiatives and dynamics as forces unto themselves.

Current scholarship has argued that within the Latin American movement toward democracy social movements do in fact "play a critical role in that struggle" (Alvarez, Dagnino and Escobar 1998). Thus, it is imperative to give an account of these social movements under new theoretical perspectives that do not treat them as mere effects or symptoms of an external cause. In this context, social movements are understood to be capable of directly causing and giving form to local social and cultural processes that can change the cultural, social and political arenas in which they act.

Drawing on Gramsci, contemporary social movements theory (especially Touraine 1985), postmodern and poststructuralist theories, and the analysis of the particularity and specificity of collective actions, some scholars began to propose new theoretical approaches to understand the complexity of social movements and their actors in Latin America. This theoretical turn is manifested in contemporary social movements' analyses. In these analyses, society is understood to be a fragmented, complex and interactive process, and social actors are viewed as multi-faceted political agents who can reshape their societies by asserting and affirming their differences. In this way, social analyses have become more complex and social actors have become agents that have political power to participate in the construction of their own social conditions and identities.

Calderón, Piscitelli and Reyna (1992) identify five approaches within Latin American studies of social movements: theories of social movements (particularly based on those developed by Alain Touraine); new forms of class analysis; analysis of communal practices; analysis focused on identity construction; and postmodern theorization that addresses complex combinations of social, economic, political and cultural causes.

Regardless of their particular theoretical approach, most current studies of Latin American social movements consider collective cultural identities and their construction essential elements in the analysis of such movements. This emphasis on culture is important in the formation and actions of social movements because it allows an understanding of the "uniqueness of Latin American social life" (Escobar and Alvarez 1992, Escobar and Alvarez 1992a).

Latin American analyses of social movements have revised the notion of social movements by taking into account their diversity and fragmentation.

> There is, then, a wide spectrum of social movements. Many of them center on specific actor, others are self-referential or monadic; some are synchronic and latent, other of long duration; some are the product of intensification of capitalism, other of exclusion; some are unprecedented, perhaps ambiguous, constantly changing, with polyvalent meanings. All of the movements, based on identities that are often changing, are internally complex and produced themselves within novel historical processes. In short, they present new historical movements in the making. (Calderón, Piscitelli and Reyna 1992:23)

Moreover, the emergence of Latin American social movements has been important to the development of theory in the United States and Europe, although such theorizing rarely addressed Latin American movements specifically in their accounts (Escobar 1992). Hellman (1992) points out how Latin American movements differ from European and North American ones: Latin American social movements arose as a result of the distrust of existing political parties where the Left was suppressed, resistance to authoritarian regimes within "imperfect democracies" and as a response to material demands and desires for personal fulfillment. In contrast, in Europe and the United States, social movements appeared within the processes of the traditional left-wing political parties' responses to postindustrial contradictions as a way to express dissatisfaction after full material needs were achieved and to overcome feelings of personal powerlessness. Slater (1988) shows how these differences are also related to welfare functions, the degree of centralization of state power, and state legitimacy.

Some scholars question the value of applying paradigms used in the United States and Europe to Latin America. However, as Escobar (1992) points out "there is no linear path between the two (multiple) places, no epistemological center and periphery . . . rather, there are multiple and mutual creations and appropriations, and resistances . . . that create an overlapping and decentered network within which both theories and theorists travel" (63). Consequently, Escobar (1992) proposes that in order to understand Latin American social movements, it is necessary to analyze the links between their economics, politics and culture from a global and interactive-process perspective. In doing that, he relates the emergence of social movements to the crisis of global development as a cultural and economic project of modernity. For him, social movements, especially grassroots movements, have the potential to postulate alternatives to western-style development, and by extension, the hierarchies and linearity of modernity itself.

> This would require changes in social relations and institutional practices, openness to other forms of knowledge and cultural manifestations—not so mediated by the language of development—and greater autonomy for communities over the creation of their own ways of thinking and doing things. (Escobar 1992:66)

Escobar argues that the crisis of modernity in Latin America has prompted the creation of new social actors and mechanisms for "the production of meanings, identities, and social relations" (68). He states that "Social movements must be seen equally and inseparably as struggles over meaning as well as material conditions, that is, as cultural struggles" (Escobar 1992:69).

Thus, analyzing cultural struggles implies understanding the daily life practices, meanings, relationships and identity construction of the social actors, because these cultural processes and modes of production constitute the basic interpretive frameworks by which people formulate and understand their social experience and construct social relations. In other words, this perspective arises from the people's self-understanding and gives them the role of agents.

Alvarez, Dagnino and Escobar (1998) have also had to expand categories and methodologies in order to account for the influence of social movements' actions in Latin America and how they change or disrupt the dominant political culture. Moreover, social movements are challenging modernity itself by proposing new conceptions of rights, economies, developments and social conditions. Thus, the cultural politics of social movements have become essential to the analyses of these movements' effects in Latin America. This "cultural politics" concept may be seen

> as the process enacted when sets of social actors shaped by, and embodying, different cultural meanings and practices come into conflict with each other. This definition of cultural politics assumes that meanings and practices—particularly those theorized as marginal, oppositional, minority, residual, emergent, alternative, dissident, and the like, all of them conceived in relation to a given dominant cultural order—can be the source of processes that must be accepted as political. (Alvarez, Dagnino and Escobar 1998:7)

From this perspective, the three factors (political opportunities, mobilizing structures, framing processes or cultural dimensions) proposed by McAdams, McCarthy and Zald (1996) have great relevance. Political actions are thus related not only to broader institutional systems but also to multiple ties that are embedded in daily-life practices. Relating these actions with the spread of meanings through informal social movements' networks expands the notion of political actions. In this way, assessing social movements' impacts implies seeing the circulation and spread of social movements' proposals, not only within the institutional framework, but also in other cultural arenas. Similarly, social movements' actions need to be assessed in relation to civil society, particularly how their cultural politics play a role in the strengthening of civil society and in the consolidation of the process of democratization which has resignified the national political culture and changed its ways of doing politics. In others words, they focus not only on "why" or "how" social movements emerge and act, but also on individual social actors and how they can act to transform their immediate conditions into democratic spaces or into a "radical and plural democracy." In this context (and for this reason), social-movements scholars in Latin America focus on the politics of the actors and their relations with processes of identity construction (Alvarez, Dagnino and Escobar 1998, Dagnino 1998, Warren 1998, 1998a).

Many contemporary analyses of collective action in Latin America focus on the process of democratization as an effect of transnational networks and grassroots organizations. The analysis of indigenous peoples' movements has been central to these studies, especially the Zapatista process in Chiapas (Urban Greg and Joel Sherzer 1991, Findji 1992, Van Cott 1994, 2000, Avirama and Márquez 1994, Brysk 1994, 1996, 2000, Varese 1995, 1996, 1996a, 1996b, Yashar 1996, 1998, 1998a, 1999, Berraondo 1999, Nash 1997, Assies et al. 1998, König et al. 1998, Warren 1998, 1998a, Ramos 1998, Gros 1998, 2000, Nelson 1999, Bengoa 2000, Laurent 2001, Ulloa 2001, Chaves 2001, Zambrano 2001, Archila and Pardo 2001, Tilley 2002, Hale 2002, Hodgson 2002, Conklin 2002, Vasco 2002, Warren and Jackson 2002, among others). However, Edelman (2001) argues that there are other movements such as those of peasants[4] and right-wing movements that are less represented in these analyses. In the following section, I provide an overview of indigenous movements in Latin America.

INDIGENOUS PEOPLES' MOVEMENTS IN LATIN AMERICA

During the 1970s, the indigenous grassroots organizations gained global recognition for political mobilizations that reconfigured their social realities. During that time, they became a valid topic of inquiry for theorists interested in collective actions. Moreover, some students of indigenous movements claim that it has been only since the 1970s that indigenous peoples' actions began to have a profound political impact within the nation-state system. Brysk (2000) claims that indigenous peoples' participation in western politics only began with the formation of the Indian Rights movement in the 1970s and 1980s. Since then, these social movements have been analyzed as political actors that have helped to change notions of identity and law by challenging traditional western conceptions of authority and order. Indigenous movements thus emerged in the 1970s as agents capable of reconfiguring national conceptions of identity, order, development, democracy and nature (Findji 1992, Brysk 1993, 1994, 1996, 2000, Avirama and Marquez 1994, Varese 1995, Pardo 1997, Gros 1998, 2000, Bengoa 2000, Ulloa 2001, among others).

In the 1970s, the idea of indigenous ethnic difference emerged as a response to emergent social, economic, cultural and political conditions. The pressures, each time more intense, over indigenous peoples' territories arose because of the opening of national borders to trade, economic growth and of the modernization processes of Latin America countries, all of which greatly impacted indigenous peoples as well. In addition, the education policies in indigenous territories, although integrative and western in orientation, helped in the formation of new leaders who developed from that educational experience concepts of ethnicity that eventually made indigenous social movements distinct from peasant movements. Practically speaking, this divergence between indigenous peoples and peasants also occurred due to struggles related to the lands they share. Consequently, indigenous activities increasingly arose from the knowledge of ethnic differences and the desire for autonomy in national contexts.

Another important condition for the emergence of indigenous movements has been the participation of internal and external agents in the processes of their construction, such as nongovernmental organizations (NGO's), international organizations (Human Rights, Amnesty International, among others), and foreign governments. The above circumstances and factors have focused indigenous movements on cultural issues and the right to ethnic difference, which has accordingly consolidated their political and organizational processes around cultural and ethnic identities (Brysk 1993, 1994, 1996, 2000, Varese 1995, Bengoa 2000).

However, some scholars, such as Bonfil Batalla (1981), González (1993), Díaz Polanco (1991), and Varese (1995), argue that the birth of indigenous movements may be traced back to the conquest of America because, from that moment, indigenous peoples began their struggle for cultural difference and autonomy. On the contrary, Bengoa (2000) argues that if they struggled, their voices were silenced by diverse processes: in the Colonial Period by racial subjection, in the Republican Period by their class subjection as servants, and in the first half of the 20th century by "*indigenists*"[5] (intellectuals, poets, artists and public employees) who took their voices and spoke on their "behalf."

Moreover, the emergence of indigenous peoples' organizations went unrecognized by some scholars, particularly orthodox left militants, because of their conceptions of indigenous peoples as culturally isolated from the main stream of historical events or as ethnic variants of the peasant class (Bonfil Batalla 1981). According to Varese (1996b) this tendency reflects a reductionist perspective or Eurocentrism that only finds relevant political actions in the indigenous movement when those actions or behaviors are recognized by the western academy according to its understanding of politics and the "course of history."

Therefore, it is important to state that throughout the colonial process, indigenous peoples did act politically through various collective social uprisings and symbolic activities (Varese 1996b). During the colonial period, in Mexico and the Andes, indigenous movements rose against the colonial power, and a series of leaders stood out who marked colonial history by expressing resistance to the new settlers. In 1780 and 1781, in Cuzco and La Paz, indigenous peoples (who recognized themselves as precolumbian Quechua and Aymara descendants) led by Tupaq Amaru and Tupaq Katari struggled against colonial power (Mallon 1992). During the 18th century, indigenous peoples participated in social and popular movements, and in this way they included the indigenous problematic within political debates of that time. Another example is the Tule (Kuna) treaty with the state of Panama in 1925, which recognized Tule culture, territory and self-determination. The Tule treaty could be considered the first political indigenous movement of this century, although some scholars regard this treaty as merely a result of US intervention (Varese 1996, Padilla 1996). It is worth noting however that the Tule people, during the colonial period, had a similar treaty with the Spanish Crown, and they established relations with the Spanish King as equals.

During the 20th century, indigenous peoples' social movement in Latin America began to assume a global political role when they participated in the First Indigenous International Meeting (Primer Congreso Indigenista Interamericano-1941) in Patzcuaro, México which tried to define the relation of state policies to the circumstances of indigenous peoples. By 1959, three more international meetings had occurred. These meetings defined the bases for the political corpus that has supported the indigenous peoples' movements in Latin America: the recognition of existence of various ethnic groups within the states, and the necessity of special policies that allow indigenous peoples to be part of the state, yet maintain their cultures (Bonfil Batalla 1981).

In the international scope, in the 1960s, the indigenous peoples of the United States, Canada, the Arctic region, Australia and New Zealand carried out public campaigns for their rights and for political recognition (Berraondo 1999). At the end of the 1960s and beginning of the 1970s, various indigenous collective actions occurred that helped to consolidate the indigenous movements and to promote their spread throughout Latin America: the Shuar Confederation against transnational oil companies in Ecuador (1964), the recuperation of indigenous lands from landlords' hands led by the *Consejo Regional Indígena del Cauca*-CRIC in Colombia (1971), and the formation of the Committee of United Peasants (*Comité Unido de Campesinos*-CUC) that brought together ethnic groups, workers, and peasants to protest against military repression in Guatemala (1978), among others. Consequently, indigenous problems, demands and political actions began to have a wider impact on national and international political arenas.

By the middle of the 1970s the indigenous movements had consolidated themselves at the global level with the creation of organizations such as the "World Council of Indigenous People" (1975) and the "International Indigenous Treaty Council (www.treatycouncil.org 1974) that grouped indigenous peoples from America, Scandinavia, Australia, and New Zealand and began to work with the United Nations" (Berraondo 1999).

During the 1980s and 1990s other indigenous movements appeared within national and transnational contexts from the *Coordinadora de las Organizaciones Indígenas de la Cuenca Amazónica*-COICA (1984), which integrates indigenous peoples from Colombia, Brasil, Bolivia, Ecuador, Perú, Venezuela, Surinam, and Guyana, to the *Ejército Zapatista de Liberación Nacional*-EZLN (1994) in Mexico; the Alliance of the Mountain Range Peoples in the Philippines (1984), the Council of the Indigenous Peoples and Tribal in India (1986), and the Asian Pact[6] (1992) (Findji 1992, González 1993, Varese 1995, 1996, Warren 1998, 1998a, Brysk 2000).

The ethnic recovery processes of the 1990s show a conceptual change from the cultural dynamics of self-preservation and survival in the previous decade insofar as they emphasize efforts to assert political power. This is evident in the commemoration preparations for the quincentennial of the Conquest of America, which generated a series of protests and reactions among the Latin American indigenous peoples who claimed a different view of the 500-year anniversary: a commemoration of indigenous peoples' resistance. It was also reflected in the demands of the 1990's indigenous insurrections such as the Ecuador Indigenous Insurrection in 1990 that took form around the claim for respect and recognition of cultural differences. In turn, this insurrection had repercussions all over America, as in the Zapatista insurrection in Chiapas that demanded indigenous recognition and autonomy and which, in turn, had a global impact on indigenous movements and political strategies around the world. In Latin America, they prompted constitutional changes which, as in the case of Colombia, caused a modification in the relations between the nation state and indigenous peoples (Bengoa 2000).

In this context, the 1990s' indigenous peoples' movements in Latin America have in common a series of characteristics: the claim that the indigenous as ethnic "peoples" differ from other citizens within nation-states; the use of ethnic identities centered in tradition as a means to promote dialogue within the discourses of modernity and human rights in order to strengthen indigenous cultural differences; the construction of new identity processes such as those related to environmental awareness; pan-indigenous relationships that articulate the interests of diverse indigenous peoples on a transnational basis; and demands for autonomy in their territories and in resource management (Bengoa 2000, Ulloa 2001).

Indigenous peoples' movements have thus been efficient in the consolidation of their political organizations, in positioning their cultural identity and in situating themselves as political actors in international and national arenas. The specific effects of their demands are evident in land recovery, in the struggles for civil and ethnic rights, in the generation of their own economic strategies, in the "conquest" of traditional political spaces by indigenous leaders, in their acceptance by nonindigenous sectors in the construction of pluriethnic and multicultural states, in the demand for new spaces of participation, and, finally, in the formation of international negotiations on a global scale (Bebbington et al. 1992).

Concerning the consolidation of the indigenous presence in the international sphere, it is important to emphasize the creation of the "Working Group on Indigenous Peoples" in the United Nations in 1982. This working group had as its background the proposal initiated by Bolivia in 1949 to create a subcommittee within the United Nations to study the situation of indigenous peoples. Another precursor important to these movements was the inclusion of indigenous peoples in the United Nations' Declaration for the Elimination of All Forms of Racial Discrimination in 1973. Also significant are the declaration of 1993 as the International Year of Indigenous Peoples, the declaration of the International Decade of the Indigenous Peoples from 1995 to 2004, and the declaration of the International Day of Indigenous Peoples (August 9).

Moreover, the report of the third "World Conference against Racism, Racial Discrimination, Xenophobia and Related Intolerance" (2001) includes the indigenous problematic and recommends an evaluation of the International Decade of the World Indigenous Peoples. It also notes the necessity of guaranteed financing for the Permanent Forum for Indigenous Issues in the United Nations system, exhorting member states to approve the United Nations' text on indigenous peoples' rights, and urging to them to reduce all forms of racial xenophobia, discrimination, and intolerance against indigenous peoples.

The formation of indigenous organizations is also the result of intellectuals (indigenous and nonindigenous), anthropologists, church members, peasants, grassroots organizations and international allies (NGOs and international governmental institutions). It is important to emphasize the role of the Barbados meeting of 1971 where a group of 15 anthropologists (most of them Latin Americans) discussed the problems of the indigenous peoples in the Amazonian region. Since then, they have formed the so-called "Barbados Group" and made the first international proclamation about the indigenous social, economic and political situation: "*Declaración de Barbados: por la liberación del indígena.*" This text had repercussions not only for some indigenous organizations, but also on academic, religious, and state institutions.

In 1977, the Barbados Group held a second meeting, also in Barbados, but at that time indigenous leaders were invited. These leaders wrote the "*Declaración de Barbados II,*" and in this case the indigenous representatives themselves called for indigenous unity in Latin America and the analysis of how indigenous peoples have been colonized, dominated and exploited. They also presented a political program for indigenous movements based on unity and the need of all indigenous peoples to change their social conditions by participating in national political arenas. They also presented political strategies: political organization according to cultural practices (a common purpose that unifies the indigenous peoples) and reaffirmation of cultural differences. They also considered promoting strategies to gain national and international support.

In 1993, in Rio de Janeiro, some of the anthropologists who participated in the Barbados meetings decided to hold a third Barbados meeting to discuss the indigenous situation in relation to the democratization of Latin America. They also produced a proclamation (*Declaración de Barbados III*) that recognized changes in indigenous social, economic and political circumstances. At the same time, they discussed how international actors (NGOs, the WB and environmental programs, among others) have created new processes of inequality. Therefore, they recognized not only the value of indigenous cultures and their autonomy, but also that of their territories and their politics of self-determination (Bonfil Batalla 1981, Grunberg 1995, Varese 1995).

Other examples of coalitions between indigenous and nonindigenous organizations are ecclesiastic groups such as *Conselho Indigenista Missionário*-CIMI in Brazil, *el Consejo Episcopal Latinoamericano* and different Catholic and Christian missions that have helped to organize Latin American indigenous meetings, to educate indigenous leaders, and to denounce the indigenous economic and social situations. Different types of

NGOs (environmentalist, human rights and legal rights) have also supported indigenous movements economically, conceptually, and politically. These alliances reflect the indigenous movements' political activism and include proposals related to citizens' rights, human rights, biodiversity, religious diversity as well as the interests of the traditional political parties (Padilla 1996, Santos 1998, Brysk 2000).

International actions such as Barbados I, II, the Congresos Indigenistas Interamericanos, the Declaration of Guadalajara (1991), the Rio de Janeiro declaration (1992), the Human Rights Inter-American commission, ILO-169 (1989),[7] the United Nations Declaration Project on Indigenous Peoples Rights (1993), the CBD (1992) and others have played important roles in the global recognition of indigenous peoples' rights and their political agency.

The indigenous peoples' movements' emergence as political agents has produced various effects, from repression to recognition. The state's violent repression of indigenous' claims was Guatemala's answer during the 1970s and 1980s (González 1993). In Peru, hundreds of indigenous people died because they were caught in the middle of the war between the state and the *Sendero Luminoso* (Varese 1995). In various countries, the rise of narcotrafficking has affected indigenous populations and spread violence throughout indigenous territories. The economic interests in biodiversity and the lack of international and national agreements to protect indigenous peoples' intellectual property rights will also remain a constant threat to indigenous peoples' autonomy. Moreover, indigenous peoples are still discriminated against racially and inequalities pervade their social, economic and political relationships with nonindigenous populations.

Nevertheless, the indigenous movements' actions have changed national constitutions throughout Latin America and have generated new processes of ethnic reaffirmation or reinvention in the region as well.[8] Some of these constitutions recognize multiculturalism or special rights for indigenous peoples; however, each constitution gives a different scope and character to indigenous rights. For example, the recognition of indigenous peoples' rights in Colombia has allowed indigenous populations, previously stigmatized, to "become Indians" again, such as the Yanacona (Zambrano 1998). Some scholars (Clifford 1988, Jackson 1991, Gros 1997, 2000, Zambrano 1998, 2001, Chaves 2001, Ulloa 2001) have shown how these processes constitute strategies of identity construction that allow indigenous peoples to relocate themselves within national and transnational spaces. Finally, indigenous peoples' movements have built a pan-indigenous identity to mobilize and center their political actions.

Indigenous peoples' movements have thus constructed a pan-indigenous identity to mobilize and centralize political actions within which many particular identities may still retain their integrity (for example: Embera, Kayapó, Tule, and Aymara). Moreover, within each ethnic group, multiple identities have emerged that cause differences and conflicts with the established political order. Since the first colonial encounters and through subsequent historical processes, both pan-indigenous and specific ethnic groups' identities have been transformed through their struggles with western conceptions of the indigenous itself. Thus the change from being perceived as an economically and socially weak person to a citizen with ethnic rights has resulted in challenges to policies or conceptions that treat the indigenous as subjects needing to be incorporated into national development programs. And, as noted above, these challenges have often resulted in the proposal and implementation of alternatives to such visions of development. Amidst all this change and conflict, the indigenous peoples' search for cultural identity and arenas for social and political expressions has been a constant. And it is this constant that provides a means to trace the changes and continuities in the historical processes that have affected the formation of indigenous peoples' movements and identities. The reality of indigenous peoples' situation in Latin America today is that they are in a permanent struggle for the recognition of their rights to self-determination, autonomy, territories, resources and knowledges.

Social movements, in general, and indigenous movements, in particular, are in a continual process of identity formation. Another way to say this is that they are continually creating new relationships to others with new meanings that become evident in the way they reinterpret norms, reshape the public discourse, challenge boundaries between public and private spaces and cultural and political fields, and prompt new ways of doing politics in which cultural differences are crucial. Struggles over meanings are thus central to these social movements and their political goals (Bonfil Batalla 1981, Cohen 1985, Touraine 1985, Escobar 1992, McAdams, McCarthy and Zald 1996, Tarrow 1998, Brysk 2000, Gros 2000, Ulloa 2001, Hodgson 2002, Conklin 2002, among others).

CONTEMPORARY INDIGENOUS PEOPLES' MOVEMENTS IN COLOMBIA

In Colombia, indigenous peoples' struggles for control over their territories began to have legal effects in the 18th century. In 1781, the indigenous peoples' actions caused the colonial government to return some of their communal territories. At the end of the 18th century and at the beginning of the 19th century, the new Latin American states, including Colombia, changed their relationships to indigenous peoples and began to consider them citizens. In Colombia, the national ideal of equality between all citizens promoted the indigenous peoples' integration and assimilation through the abolition of tribute, the institution of monetary payment for work and the privatization of land in order to convert the indigenous into property owners.

Such notions of equality and citizenship required that *resguardos*[9] be dissolved. Indigenous peoples often opposed this, claiming their original rights under colonialism. These policies generated diverse processes at the national level. In the north, *resguardos* were effectively dissolved, while in the southwest there was much resistance to the process. Also, privatizing property did not generate an increase of individual owners; rather indigenous peoples were assimilated as groups into the latifundium and they became servants or *terrazgueros*.[10] In 1837, indigenous peoples opposed and actively protested the dissolution of their collective territory.

Under these social circumstances the national government promulgated laws to protect indigenous peoples. Law 89 of 1890 promoted special treatment for the indigenous peoples in their territories so that they could become modern. Through this law the indigenous territories were declared the property of indigenous peoples and their local authorities and councils gained legal recognition. Even though the law's original purpose was to civilize indigenous peoples and transform them from "savages" to modern citizens, during the last 100 years indigenous peoples have used it to recover their ancestral territories, to maintain their cultural practices and to keep their political authority in councils.

At the beginning of the 20th century, indigenous leaders such as Manuel Quintín Lame, José Gonzalo Sánchez, and Eutiquio Timoté participated in the national political arenas, and even though they had different political interests, they constructed the common basis of subsequent indigenous movements: a sense of indigenous cultural and political distinctiveness. But only at the beginning of the 1970s with the appearance of the first indigenous organization, the Cauca Indigenous Regional Council (*Consejo Regional Indígena del Cauca*-CRIC 1971), were indigenous peoples able to take part in national politics through their own organizations. These organizations arose from a "discourse of ethnic difference" that the indigenous peoples had struggled to insert within the state and national political processes in order to gain national recognition of their rights and differences. By doing so, indigenous peoples asserted their rights as "the legitimate owners of America" and militated for legislation to support indigenous peoples in their efforts to recover their traditional territories and defend their cultural heritage. Their actions promoted the construction of new political relations and the power to negotiate with the state, the private sector, other indigenous groups, and social movements as well as paramilitary and guerilla organizations.

After the rise of diverse regional and local organizations, the First Indigenous Regional Meeting was held in Tolima, Colombia in 1974. Later, in 1980, another meeting was held in the Tolima department called "The First National Meeting of Lomas de Hilarco." One of the results of these meetings was the idea to create a national indigenous organization. In 1982 the first indigenous national organization of Colombia called Organizacion Nacional Indígena-ONIC emerged in the national political context. Thus, over time, the indigenous peoples and their organizations consolidated a pan-ethnic movement with clear purposes based on their demands for the recognition of the cultural and ethnic diversity within the Colombian state, the autonomy to control their territories and natural resources, and the defense of their traditions.

At that time, indigenous peoples' movements aligned themselves against the national governmental program called "The Indigenous Statute" (*El estatuto indígena*) that proposed changes in the previous legislation related to indigenous peoples' territories in order to dissolve their communal territories, to reform the political councils and to end collective property. Indigenous peoples opposed this proposal and used Law 89 to keep land tenure a matter of collective property. Their position often generated repression of their leaders (Gros 1991, Ramírez 1994, Laurent 2001). Since then, different indigenous organizations have been formed at the local, regional, and national level.[11]

Indigenous peoples' organizations have used a variety of means to promote their various interests and goals, from negotiation with the state to the formation of armed organizations such as the Armed Movement "Quintín Lame" (1974). This variety arises from the fact that these different organizations have had diverse origins, political strategies, political interests, identities and territorial concerns. Also, these organizations have had support from various religious, social and political organizations: the peasant movements, workers' unions, intellectuals, and leftist activists, among others (Findji 1992, Avirama and Marquez 1994, Laurent 2001).

The indigenous movements' political actions have helped to change Colombian national policies and have created special programs within the state that have allowed more participation for indigenous peoples and some degree of autonomy in their territories. In 1980, the *Programa de Desarrollo Indígena-PRODEIN* was created in an effort to incorporate indigenous proposals as well as resolve indigenous socio-economic problems. Policies of integration were thus changed into policies of participation. After that, the state legally recognized different indigenous territories and created state institutions (for example, the Ministerio del Medio Ambiente=MMA and Instituto Colombiano de la Reforma Agraria-INCORA) and programs to negotiate with indigenous peoples. Moreover, in 1991, the state ratified ILO-169 through Law 21 which recognized indigenous populations as "peoples" with the right to maintain their cultural identities and territories. However, it was only with Colombia's National Constitution of 1991 that indigenous peoples were recognized as full citizens (Gros 1991, 2000, Ramírez 1994, Laurent 2001).

In the 1980s, strong indigenous peoples' movements emerged in Latin America and Colombia and gained recognition within national and international political arenas. In Bebbington et al., terms (1992), we can speak of "The Found Decade" because the indigenous peoples consolidated their organizations, positioned their cultural identity and thereby constituted themselves as social actors and protagonists in political and social processes in the national and international arenas.

The presence and actions of these social actors and movements cannot be separated from international transformations that began in the 1970s. Globalization and democratization processes, linked to the spread of technology and communications, reconfigured global-local processes and changed the spatial and temporal situations of the nation-state and social movements, by articulating them within transnational space. In Colombia, the indigenous recognition process obtained through the Colombian Political Constitution of 1991(CPC-91) was linked to the transformation of the state by decentralization and the implementation of neoliberal policies (including the privatization of the state institutions, the abolition of subsidies, and the opening of the Colombian market to the international market). There was also an emphasis on discovering and eliminating the discrimination evident in policies affecting indigenous peoples (Gros 2000).

The National Constituent Assembly

At the beginning of 1980s, the Colombian state created an administrative decentralization process and a constitutional review through the National Constituent Assembly that it consolidated in 1991 with the promulgation of the New National Political Constitution, hereafter CPC-91. In this process, indigenous peoples' participation became very important in national reconstruction (Gros 1991, Ramírez 1994, Laurent 2001).

Avirama and Márquez (1994) describe the process in the following way:

> Thus arrived the elections for the National Constituent Assembly (Asamblea Nacional Constituyente), during which the Indians achieved the creation of a democratic block, made up of their three delegates—two through election and one through the peace process—and the political parties Democratic Alliance M-19 (Alianza Democrática M-19) and National Salvation (Salvación Nacional). Out of the Constituent Assembly came a new constitution, in which for the first time rights were assigned to the indigenous peoples and communities with respect to territory, politics, economic development, administration, and social and cultural rights. (1994)

The indigenous peoples' representatives pursued discourses and proposals on two topics: their own needs (political, administrative and territorial autonomies or sovereignty over their territories and resources); and the social, economic and political needs of the nation as a whole. In this way, the indigenous participation made evident that Colombia was a nation of multicultural and pluriethnic character. On this basis, they made political claims regarding ethnic groups' autonomy (not only indigenous peoples but also Afro-Colombian communities), the necessity of delimitation of ethnic territories and administrative order (in agreement with the sociocultural reality) and, finally, a special method of participation within the electoral system for indigenous peoples. Likewise, emphasis was placed on the discrimination that indigenous peoples had been subjected to historically. The indigenous presence at the constitutional proceedings also produced discussion about nature and the environment and calls for a cessation of contamination of the rivers and the atmosphere (Laurent 2001).

CPN-91 legally recognized the nation's cultural and ethnic diversity. Under this constitution, indigenous and ethnic groups have gained recognition as citizens with special cultural rights. These cultural rights allow indigenous peoples self-determination and their own government and autonomy in their territories, which implies decision-making in an autonomous way in the administration of justice and control of territory as well as resource management (Articles: 7, 8, 10, 19, 63, 68, 70, 72, 96, 246, 329 and 330). Indigenous peoples also have a special electoral jurisdiction that allows them to have two representatives in the Senate. Also, this decentralization implied a territorial reorganization according to the regional, economic, cultural and historical diversity of the country. The new Constitution also created a new political arena, the Indigenous Territorial Entities (*Entidades Ter-*

ritoriales Indígenas-ETIS) (See articles 286, 287, 329 and 330) (Ramírez 1994, Dover and Rappaport 1996).

According to the CPN-91 (article 330), the indigenous territorial entities should be governed by their own authorities and have the autonomy to "design the policies, plans and social, economic development within their territory, in harmony with the National Development Plan; to promote public investment in their territories and surveillance of their due execution, to perceive and distribute their resources; to watch over the preservation of natural resources, among others." Also, the same article established that "natural resources exploitation in the indigenous territories will be done without damaging the economic, social and cultural integrity of the indigenous communities. In the decisions adopted concerning such exploitation, the government will provide for the participation of the respective communities."

For indigenous peoples this article means that the Councils will govern their territories or ETIS according to indigenous cultural practices. In the places where indigenous peoples are the majority, the councils will be mainly composed of indigenous members; as a consequence, indigenous peoples will govern nonindigenous people for the first time in Colombian history (Ramírez 1994). According to Dover and Rappaport (1996), the consolidation of these ETIS implies a new form of citizenship. However, this article has not been implemented yet.

After the Constitution of 1991: Indigenous Peoples' Political-electoral Process

After the constitution, and thanks to a special electoral political arena, indigenous peoples' political organizations began their political-electoral participation. This electoral arena admitted diverse representatives to promote the coalition and cooperation of the indigenous' organizations with diverse nonindigenous political parties. However, this arena has also generated divisions, differences, and heterogeneous opinions in the electoral-political field.

In the 1990s the main indigenous peoples' electoral-political organizations appeared and consolidated themselves as such: The Indigenous Social Alliance (La Alianza Social Indígena-ASI) (1991), the Colombian Indigenous Movement (Movimiento Indígena Colombiano-MIC) (1993) and the Indigenous Authorities Movement of Colombia (Movimiento de Autoridades Indígenas de Colombia-AICO) (1994). The creation of these organizations and the diversity of their candidates, proposals and interests revealed changes vis a vis the interests and proposals of the previous decades' indigenous organizations. This also revealed the internal divisions of the diverse indigenous organizations. Also, the actions of these organizations have been differentiated according to the influence of their regions and their alliances with established political parties. Some have not obtained enough votes to gain a political position while others have positioned several of their representatives in the congress, mayoral offices and departmental governments through the support of indigenous and nonindigenous voters as well (Laurent 2001).

Laurent (2001), analyzing the electoral political panorama of the last ten years, establishes a difference according to two periods. From 1991–1994, or the first period, in spite of indigenous positioning in the national electoral sphere, their political presence was marginal in comparison with the traditional parties. There was a decrease in votes in areas of indigenous peoples' parties, which evidenced low support of indigenous peoples for their candidates. In contrast, in the urban areas, votes in favor of indigenous peoples were higher. For this reason, Laurent argues that during this period, the elected indigenous representatives, mainly for the Congress, gained offices in great part from the nonindigenous sectors' voters who found those candidates an option to the traditional political parties, a way to protest against national policies, and a way to provide support to the indigenous cause. However, at the regional level, for departmental and municipal elections, although the indigenous voters participated, the multiple indigenous lists fragmented their votes. These results generated changes in the political strategies of the indigenous organizations, as reflected in the following electoral period. In the second period, in the departmental and municipal elections of 1997 and in the parliamentary elections of 1998, the numbers of elected indigenous representatives increased, but the problems of the multiple lists remained, as did the differences between the organizations.

For the period between 2001 and 2002, indigenous representation increased and made even more evident the diversity of the lists and prompted the creation of new political parties and alliances with other nonindigenous parties.

Indigenous electoral campaign mechanisms are the same as those used by traditional politics: posters, banners, brochures, and advertising through mass communication. However, because the nonindigenous electorate is very important in the indigenous political process, it plays an important role in the creation of "indigenous" images. The main topics that appeal to nonindigenous voters are ecology, the defense of the environment, the maintenance of indigenous peoples' ancestral knowledges, social justice, subsistence-based economies, and intercultural relationships, all of which fall under the main idea of constructing a socially and ecologically respectful nation that values cultural diversity. Consequently, the images used in electoral campaigns keep certain symbolic and ethnic markers of difference such as the dress and the long hair that evoke the cultural imaginary of "indigenousness" in national society (Laurent 2001).

Indigenous peoples have "conquered" political arenas in the national congress, mayoral offices, and departmental governments, as in the case of Floro Tunubalá who is a Guambiano and was governor of Cauca, Colombia (2001–2003). Indigenous participation in the electoral-political scenario has not only given a voice to indigenous peoples in political discussions that affect projects and laws in the indigenous territories, but also in discussions that affect the whole nation. In particular, there has been discussion of problems such as those of the Uwa[12] or Embera-katio[13] peoples' influence over national natural resource policy. In this way, interaction between national purposes and indigenous peoples' movements has favored the indigenous. However, indigenous peoples' presence in arenas of political power has also produced conflicts with nonindigenous participants and traditional political parties. Traditional parties often try to limit indigenous participation because of indigenous people's inexperience in those arenas or because the indigenous confront them and their interests when efforts are made to weaken national laws that promote indigenous peoples' interests and autonomy.

In some instances, the political process has also led to indigenous representatives' alliances with traditional political parties and bureaucratic machinery that can delay or transform their political rights through co-optation. Jackson (1996) points out that in this process the "Indian organizations become, in many ways, agents of the state, with similar bureaucracies, language, constructions of what needs to be done and how to do it" (140).

Also, the diversity of indigenous organizations and their positions may impede unified proposals to address the indigenous peoples' demands within traditional electoral politics. Furthermore, within the indigenous movements there are critiques of the aftermath of the 1991 constitution. These critiques focus on indigenous representatives to the senate, mayoral offices and departmental governments who have made political alliances with the traditional political parties in ways that may have negative implications for indigenous interests. In addition, there are criticisms of the distance of these representatives from the local indigenous communities and the lack of grassroots support for them.

Finding a political balance after more than ten years under the CPN-91 has not been simple for indigenous peoples. Their presence in the electoral political process has given them a higher visibility at the national level and entry into political arenas with a potential to act, yet there are still problems arising from their complicated relations with powerful nonindigenous interests. In addition, they have experienced many divisions amongst themselves, a growing awareness of their lack of training in official politics, and general confusion among their representatives, organizations, and communities (Laurent 2001).

Nonetheless, indigenous movements, with or without their political representatives, continue the strategies that have been used since the 1970s: occupations, interruptions, direct actions, civil disobedience, international alliances, and political mobilization, among others. The last Colombian Indigenous Peoples Congress in November 2001 proposed a national statement to coordinate indigenous purposes and interests related to territorial autonomy, development models and the peace process. This involved the cultural, social and territorial reconfiguration, not only of indigenous territories, but also of the country as a whole under the principles of regionality, cultural identity, diversity, cultural autonomy, sustainability and political participation. Colombian indigenous peoples put forward the idea that respect for diversity and differences will consolidate the peace process and open space for a new economy that guarantees "the internal production of the basic Colombian family's expenses, the right for food of the whole nation and the rehabilitation of the countryside from the effects of export policies, big transgenic food importation and the stealing of peasants' lands" (www.onic.org.co).

With respect to their territories, the indigenous peoples' demands are based on previous recognition of indigenous territories by the state: cultural, political, administrative and budgeting autonomy in the management of their own interests; the right to be governed by their own authorities; and participation in managing national economic resources and the administration of taxes. This has made it possible for them to take part in the nation's cultural, environmental, social, economic and political life while preserving the collective patrimony of indigenous peoples, which includes their own existence, their cultural heritage, their traditional environmental knowledge, and the biodiversity and genetic resources that exist in their territories (www.onic.org.co).

With respect to models of development, they propose "an economy that has three principles: to protect nature, to guarantee that collective interests prevail over entrepreneurial or private interests, and to guarantee food security. All of these promote recognition, respect, and support for economic models of collective solidarity, not neoliberal models, and political reforms to institutionalize social control over the great economic topics" (www.onic.org.co).

In the context of current political conditions and the civil conflict at the national level, indigenous peoples find themselves in the middle of a series of new processes related to military confrontations in their territories and displacement and concentration in urban centers. In response, indigenous peoples are proposing actions to protect their way of life, dignity, territories and resources through the spirit of resistance and peaceful actions, as well as the creation of special zones that allow them more social control and territory.

Indigenous peoples' protests related to environmental concerns have also been expressed in vetoes of anthropological and biological research on genetics that have not received their prior informed consent. In 1996, the ONIC and diverse indigenous peoples' organizations made a declaration against such research programs and the taking of genetic samples of human populations carried out by the University Javeriana. In addition, in 1999, COICA achieved the repeal of the patent on Yagé that had been granted to a North American researcher.

In April 2000, 5000 delegates of 48 ethnic groups participated in a national mobilization to support the Uwa and Embera-katio. This mobilization involved different political expressions—from blockades of roads to a hunger strike by some of the indigenous peoples' congress members—to press the government concerning the problem of these two cultures. Peasants, fishermen, civil society, and international and national NGOs supported them.

Indigenous peoples' coalitions with regional, national and international actors have been successful as a means to carry out their protests. These are important because they promote indigenous interests by creating juridical leverage in the recovery of lands, cultural recognition, and political organization and autonomy. They also appeal to solidarity with transnational civil society and environmental and human rights NGOs in order to create international pressure on the national government. They also request foreigners' intervention as observers of political processes and as enforcers of the agreements. (As noted before, environmental and human rights NGOs have intervened to support the Embera-katio and Uwa people in their causes.)

In this context, indigenous peoples' actions related to the ethnic and cultural demands that the CPN-91 recognized stand out. These demands address how to promote indigenous peoples' own forms of justice and administer it in an autonomous way in accordance with the use of their Law (called the Indigenous Peoples' Special Jurisdiction), how to participate in different political processes (National Plan of Culture, Process of Peace and national policies of education among others), and how to regulate the ETIS and control their own natural resources.

In spite of CPN-91's recognition, indigenous peoples are in a constant struggle to promote their issues, especially autonomy and self-determination, through national institutions and in a civil society often consumed by military issues and guerilla conflicts. Therefore, although they have struggled for the recognition of territorial rights of the Embera-katio and Uwa peoples, for example, and they are constantly denouncing the violation of human rights (murders, kidnappings, disappearance of leaders, and forced displacements, among other things), they maintain neutrality in the face of armed conflict.

THE CULTURAL POLITICS OF INDIGENOUS PEOPLES' MOVEMENTS

After many years of contesting their marginal position through political actions, indigenous peoples' movements have finally located their proposals within the national project. In this way, indigenous peoples have helped to rethink predominant notions of citizenship, development and democracy. Indigenous peoples' proposals are thus based not only on the recognition of their specific cultural differences, but also on the more general conceptual distinction between communitarian and individualistic conceptions of citizenship and politics. Consequently, they claim collective rights over their territories, and they defend their communal organization under *cabildo* jurisdiction, as in the Colombian case. Moreover, they have located their struggles for their territories within historical processes that relate their precolonial situation and values to current political processes.

Indigenous peoples have based their demands for rights to self-representation and sovereignty over their territories on their cultural identities. Thus, indigenous movements define their members as *indigenous peoples* rather than ethnic or racial minorities within a dominant national society. Consequently, indigenous peoples call themselves 'original people"[14] (legitimate territorial owners) or nations that demand restitution rights and ancestral sovereignty in their territories. Moreover, indigenous peoples establish political relationships with the state from this perspective and through this authority. In this way, they reaffirm their autonomy and self-determination while they recognize the state's institutions. Indigenous peoples' movements thus demand a national understanding based on the recognition of their difference as peoples (Rappaport and Dover 1996). This assumption manifested itself in one indigenous movement's campaign slogan for the National Constituent Assembly: "Because we defend our rights. We support the rights of everybody. Vote for the Indigenous Authorities candidate" (Findji 1992).

According to Findji (1992), the indigenous peoples' fights for rights have initiated a new way of doing politics:

> In the cultural order that surrounded the shaping of Colombian society, rights for most people have been precarious, and public discourse about them has been quite limited. To listen to the indigenous peoples demanding their rights immediately suggests another cultural order—a long, collective memory with which most Colombians, a relatively new people of colonizers of unused public lands, were unfamiliar (125).

Indigenous peoples' movements participate in the process of building new forms of democracy through civil practices that expand the idea of citizens' rights. Moreover, these movements have created spaces for new kinds of social relationships and political actions, and their fights have led to the definition of national political practices that propose to establish relations with the state based on conceptions of reciprocity. Thus, indigenous peoples have also used their claims for rights as indigenous peoples to expand democracy within the nation-state. Indigenous peoples have thus become political actors whose agency allows them to build civil practices that can transform the modern notion of the nation-state. Indigenous peoples' movements consequently demand not only democratic spaces but also redefinition of the ideas of rights, equality, difference, individuality and collectivity, among other concepts. As a result, indigenous movements have acquired a recognition and power in the dominant political culture that help them to redefine that dominant political culture.

In Colombia, indigenous peoples have to be recognized under the new national constitution as different and as citizens with distinctive ethnic rights. Thus, they claim cultural differences based on language, law, life conceptions, development and particular relations to their environment. Therefore, indigenous peoples are recognized under the idea of "nation" (unique tradition, identity, law, language and collective territory) within the state. In this way, indigenous peoples have constructed themselves as a "collective identity." Consequently, indigenous peoples have to become "traditional" in order to be included in the national arenas. In Gros' (1998) words, they are in the situation of "being different in order to be modern (the paradox of identity)."

This process, in Comaroff and Comaroff's (1997) perspective, could be seen as a result of the indigenous movements' agency. Following their definition, the indigenous movements' agency causes meaningful consequences and articulates new discourses of representation in the nation-state. In addition, indigenous movements' actions contest the law by means of the law by negotiating and relocating themselves within the national constitution

(Lazarus-Black and Hirsch 1994). In this way, they "use" their collective identity as a performance strategy in order to establish relations with the state (Gros 1998, Bourdieu 1977). In Colombia and elsewhere, it is this political strategy that allows them to manipulate their cultural and historical situation. In this way, "collective identity" becomes an ongoing historical construction that allows the indigenous movements to struggle for their political and practical interests within the national and international arenas. Consequently, these minorities have, in Gros' terms, an "open ethnicity," which implies the possibility of new, flexible conceptions of indigenous identity that allows them to deal with the contradictions that follow from their entry into modernity and the nation-state. In Bourdieu's terms, this shift could be the beginning of the transformation of the "official strategies".

Thus we can say that the collective identity of indigenous peoples' movements originated in the three sources noted by Castells (1997). One is the way indigenous peoples were recognized as ethnic actors by a dominant state institution, another is the national constitution which recognized them as indigenous peoples (in Castells' terms, a *legitimizing identity),* and the third is the indigenous' political struggle since the 1970s that has helped to define their identity as a result of processes of resistance (a *resistance identity).* Finally, since the new constitution, the new indigenous movements have built for themselves new identities based on ethnic traditions in relation to transnational discourses of ecology, cultural diversity, alternative development and human rights that are different from the national identity (*a project identity).* Therefore, indigenous movements as collective identities have created a political presence that has entered the established political arena to challenge the official system. Moreover, indigenous movements have "manipulated" the legal system by not only using it, but also redefining it. In this way, following Castells' (1997) ideas of social movements, the indigenous movements' actions transform "the values and institutions of society" (3). Nonetheless, according to Rappaport and Dover (1996), members of the different indigenous organizations may also construct their differences based on traditions and anthropological and legal definitions of *indianness* that reflect national stereotypes. In this sense, the identities that they project and the goals that they set also change according to the indigenous organizations' relations with the dominant groups' political, social and cultural interests.

FINAL REMARKS: INDIGENOUS PEOPLES' POLITICS, NEW RECOGNITION?

Indigenous peoples' movements have consolidated new ways of doing politics and have proposed their knowledge as an alternative for managing natural resources. However, ever since the colonial period such equality has not arisen simply from the recognition of difference. Indigenous peoples are still without clear access to many social and political rights. Similarly, indigenous peoples are still displaced from their territories because of violence, drug traffickers, paramilitaries and guerrillas. Moreover, poverty still is one of the most important problems that they face. In addition, indigenous leaders have been persecuted violently by large landowners (Avirama and Márquez 1994). Furthermore, there are many internal divisions within indigenous organizations that have led to political competition for national and international financial resources. Thus indigenous movements are still struggling for participation in political affairs as part of their effort to gain recognition of their difference and for their proposals for social, environmental and economic development.

Nevertheless, the new political context in Colombia, for instance, does allow the cultural politics of the indigenous peoples' movements to open spaces for cultural differences and ethnic rights within the national political arena. This is a new process in Colombia because, for the first time, the indigenous peoples are in a position that allows them to have and exercise political, civil, social and cultural rights. For example, in Colombia, indigenous peoples have gained new rights over their territories and natural resources that allow them to propose alternative developments based on their practices. In this way, indigenous peoples are enjoying formal citizenship, and they are opening arenas for the recognition and implementation of their substantive citizenship. Accompanying these accomplishments is greater national and international recognition of indigenous peoples' practices and knowledges. According to the history and events described above, indigenous peoples now have unprecedented circumstances for situating themselves as powerful political actors within their nations and the global eco-political arena. In turn, this recognition has required a shift in modern conceptions of democracy in order to accommodate indigenous political rights and cultural differences as well as their conceptions of their relation to nature.

In the last decades, some states have recognized multiculturalism or special rights for ethnic groups, which seem to indicate that multiculturalism and the recognition of difference can be seen as a result of the novel emergence of collective identitarian actions. However, some scholars claim that multiculturalism is not a new element in the process of constructing nationalism. They argue that diversity has been part of the idea of nation at different historical moments (Bhabha 1994, Wade, 1997). Bhabha (1994) describes how different narratives can coexist within the nation: "The people are not simply historical events or parts of a patriotic body politic. They are also a complex rhetorical strategy of social reference: their claim to be representative provokes a crisis within the process of signification and discursive address" (145). Moreover, minorities and their political struggles and resistance to the western conception of the nation have always been part of the complexity of that conception of the nation, and they are necessary for the constitution of the western nation as the ethnic Other that is necessary to distinguish the nation's own identity and that forms the measure of its continuity and change. In this way, the modern nation not only constrains diversity but also produces and maintains it and, sometimes, changes in response to it.

According to Collier, Maurer and Suárez-Navaz (1995), this recognition of difference is part of the "bourgeois law," which "has the major role in producing such differences." They consider that "bourgeois law" produces difference in two ways. "First, by declaring everyone equal before the law, it constructs a realm outside of law where inequality flourishes. The ideal of equal treatment before the law not only makes it difficult for law to address, and thus to redress, the differences in power and privilege that law defines as occurring outside of or before it, but legal processes actually enforce and confirm inequalities among people in the processes. Second, bourgeois law demands difference even as it disclaims it, both soliciting expressions of difference and enforcing the right of people to expresses their differences even as law requires people to stress their similarities in order to enjoy equality" (2).

Other scholars claim that the inclusion of minorities is only a new way of controlling people who could not otherwise be integrated into the nation-state. Consequently, they argue that these changes seem to continue the unique vision of liberal law. Wade (1997) notes that strategies of recognition "often seem to obey motives of political control, and this indicates that these new trends are still subject to the play of power and resources" (1997:105).

According to Hernández and Ortiz (1994), when the legal system recognizes the indigenous, it does so because the political struggles of indigenous movements confirm its own power and help it create new forms of control. Other scholars argue that when the state is defined as multicultural and multiethnic, the state may be considered a "part of an increasingly global discourse of cultural identities and the particular position of the nation-state in the larger political economy of such discourse" (Motzafi-Haller 1995) does not necessarily operate only to confirm its own power.

Gros (1998:32), analyzing Colombia, claims that the state constructs different identities because it "needs ethnic actors, well-defined, recognized and legitimate, in order to negotiate its own intervention." Thus, when the state creates a general idea of "collective identity"—the other as indigenous—it often does so with the goal of identifying the individuals under that identity as people of special interest to the state insofar as they pose "problems" to its norms and functions. Moreover, the state gives them the basic conditions for their organization (legal recognition, economic resources, just to name a few) in order to "mediate with all its power but with a new language, [in order] to pervade the communities with its rationality and instrumental modernity."

As noted before, Hale (2002) states that the rise of multiculturalism is also related to neoliberal policies that embrace the right of ethnic recognition. "The state does not merely 'recognize' community, civil society, indigenous culture and the like, but actively, re-constitutes them in its own image, sheering them of radical excess, inciting them to do the work of subject-formation that otherwise would fall to the state itself" (496)

Other scholars (Sierra 1990, Hernández and Ortiz 1994, Ortiz and Hernández 1996, Nelson 1999) note that indigenous rights and the conformation of pan-indigenous identity have also generated conflicts within and between indigenous and peasant communities. Ortiz and Hernandez (1996) point out how, in Mexico, indigenous and mestizo women have contested generic notions of indigenous rights because these pan-ethnic identities do not consider, for example, particular conceptions and situations of women within indigenous communities. In a similar way, indigenous women criticize ethnic cultural revivals that give indigenous traditions a status of moral superiority by questioning the inequalities that mark their daily lives within those cultures.

Edelman (2001) also calls attention to the problems of identity-based mobilizations, noting that they constitute opportunities to gain political participation while, at the same time, they also pose risks of political fragmentation for emerging social groups. He states that "claims of difference can fortify demands for new rights, but they can imply an abdication of rights as well" (299). Finally, Edelman following Klein (1999:115), calls attention to the effects of defining identity in the context of how corporations and states have promoted "'diversity' as 'the mantra of global capital,' and used it to absorb identity imagery of all kinds in order to peddle 'mono-multiculturalism' across myriad differentiated markets" (300).

Within the social-movements theories there is also the recognition of the indigenous peoples' autonomy as one of the successes of indigenous movements and their particular identities, and some scholars argue that this recognition is new. However as Colchester states:

> Some of the jurisprudence that underpins the contemporary recognition of indigenous peoples is as old as the history of conquest. Conquering powers since the Romans have recognized that native peoples should enjoy some measure of self-governance and their rights to exercise their customary laws. Policies of 'indirect rule' were also widely favoured by the British and Dutch in their colonies. It seems likely that such policies were adopted not so much out of respect for cultural differences, but rather as the least contentious and cheapest way of maintaining imperial control [. . .]. Nonetheless colonial laws, to a surprising extent, affirmed the principles that native peoples have the right to apply customary law and represent themselves through their own institutions. In the 18th and 19th centuries, the colonial powers did not hesitate to deal with native peoples as 'nations' and to sign treaties with them—often with the aim of cheating them out their sovereignty and lands, is true. But these legal precedents have provided the basis for the emergence of new jurisprudence which establishes current notions of 'aboriginal rights' and doctrines of legal pluralism (Colchester 2002:2).

In the last decades, the indigenous peoples' identity construction process has become linked to western ecological ideas, and indigenous peoples have become participants in a national and transnational dialogue that employs discourses that address environmental changes primarily in western terms. This latest phase of indigenous identity construction is clearly related to the rise of environmental awareness in the West of the differing ecological practices around the world. Accordingly, the indigenous peoples' practices, conceptions and knowledges about humans and nonhumans, nature and culture, have influenced western thinking at the national and international levels. In particular, nonindigenous national and international NGOs have promoted the recognition of the indigenous peoples' conceptions of nature. In this way, an ecological identity has been conferred on indigenous peoples who, at the same time, have contributed to the existence of that identity by reaffirming their identity, practices and conceptions in its terms. In this sense, they have contributed to their own status as "green bodies" within the western ecological imaginary.

Chapter Three
Indigenous Peoples' Movements of the Sierra Nevada de Santa Marta, Colombia

INTRODUCTION

The previous chapter formed the general framework needed to locate and focus this chapter on the emergence, development and consolidation of the political and social movements of the inhabitants of the SNSM: the Kogui, Arhuaco, Wiwa, and Kankuamo[1] peoples. Their cultural and environmental proposals have reached longstanding national and transnational political arenas where they have created a dialogue with both governmental institutions and NGOs. This dynamic interaction of culture, environment, organizational processes and politics has also led to the formation of new public arenas in which the indigenous peoples of the SNSM have expressed their political values and goals.

As discussed earlier, the organizational processes and political struggles of indigenous peoples of the SNSM have developed through conflict and cooperation with transnational indigenous peoples' political organizations and the national governmental institutions of Latin America and Colombia in particular. While Colombia's national indigenous organizations have not fully coordinated themselves in these national and transnational contexts in a permanent or sustained manner, they nevertheless share political struggles, mutual problems and interests with them. In this chapter, I address the CTC's process of organization, and how its proposals represent an indigenous environmental and territorial knowledge that incorporates the cultural values and practices of the SNSM's peoples and that permeates the policies of a series of local, regional, national, and transnational actors (including mayor's offices, corporations, ENGOs, state institutions and the World Bank). These interactions between the CTC and the state institutions have established agreements, revealed differences and built new strategies regarding territorial, cultural and environmental issues, thereby empowering the indigenous peoples of the SNSM in their efforts to propose new and different territorial and environmental relationships with nonindigenous peoples.

Indigenous peoples' cultural politics form the basic ethnographic context for the analyses in the following chapters. Using documentary evidence from indigenous peoples' movements, I demonstrate the linkage between their cultural politics, collective identities, territorial sovereignty and political goals of self-determination and autonomy. The following chapters will then address how indigenous peoples' environmental and cultural politics have often produced varying and often contradictory effects within national and transnational political arenas, especially with respect to global ecopolitics.

INDIGENOUS PEOPLES' ORGANIZATIONAL AND MOBILIZATION PROCESS IN THE SNSM

In the Arhuaco zone, indigenous grassroots organizational processes began in the 1930's. The Arhuaco people started a political organizational process in 1931 when they formed the Indigenous League of the Sierra Nevada, which was linked to the Workers Federation of Magdalena. Later, in 1974, the Arhuaco Indigenous Organization and Council (Consejo y Organización Indígena Arhuaca-COIA[2]) was formed to recover territories and to gain recognition and respect for Arhuaco culture. This organization was subsumed and succeeded by another organization called the Tairona Indigenous Confederation (Confederación Indígena Tairona-CIT) (Uribe 1993).

From 1974 to 1987, the CIT was the only indigenous organization located in SNSM. This organization now shares strategies and political processes with several national movements, as in the case of the ASI, and occasionally it has participated in processes generated by the ONIC. During the last 15 years other indigenous organizations have been established, which include: Gonawindúa Tayrona Organization (Organización Gonawindúa Tayrona-OGT), Organization Wiwa Yugumaiun Bunkwanarrwa Tayrona (Organización Wiwa Yugumaiun Bunkwanarrwa Tayrona-OWYBT), Kankuamo Indigenous Organization (Organización Indígena Kankuama-OIK), and the Organization Avimolkueise (which disappeared in 1995).

Since 1999, the CTC has been consolidated in an effort to form an overarching organization to coordinate the grassroots organizations of the four indigenous cultures of the SNSM. Below I describe the CTC's formation process and its political negotiating strategies, which requires a historical description of the OGT's development and activities and how they led to the CTC's formation.

Gonawindúa Tayrona Organization (Organización Gonawindúa Tayrona-OGT)

The OGT organization emerged in 1987 as a response to the needs of Kogui, Wiwa, and Arhuaco peoples who live on the northern slope of the SNSM. Kogui, Wiwa, and Arhuaco peoples and their *Mamas*[3] who recognized the need for a grassroots organization able to interact with state institutions in the centralized and discursive manners those institutions require to fulfill their purposes. Consequently, they designed this indigenous organization to interact with nonindigenous agents and institutions in accordance with the principles of cooperation and participation formulated in The Original Law[4] and evident in the indigenous peoples' culture and traditional practices. The *Mamas* saw such an organization as necessary for effective interaction with the outer world, especially with respect to their efforts to recover their territory and stop unsupervised national and local interventions. This objective was synthesized in the organization's name and goals. Even though the CIT already represented the problems and needs of the SNSM's indigenous peoples, the three indigenous groups of the northern slope did not consider themselves represented adequately by that organization. This was due to both spatial and conceptual distance: CIT had not come from their territories or The Original Law. The Original Law forms the basis of the SNSM indigenous peoples' thinking, and with the formation of the OGT it appears as the simultaneously cultural and political argument upon which they founded their claims to territorial restitution and the right of political self-determination (Rubio 1997, OGT 1998).

Consequently, the OGT'S formation had the particular distinction of being legitimized by the *Mamas* of 21 different communities.[5] The *Mamas,* through divination processes, pagamentos,[6] and debates in the nuhue,[7] recognized the OGT's formation as consistent with the Original Law. According to the *Mamas,* this process reasserts the internal decision-making processes and organizes the two aspects that configure Kogui thinking: the spiritual and material planes. Although the *Mamas* were the organization's driving force, people from governmental institutions also supported the OGT (for example, the head of national Indigenous Affairs Office was important in its formation).

The organization's spiritual character arises from the *elder brothers*[8] responsibility to care for The World's Heart by maintaining its harmony and balance. Their goal is the continuity of the indigenous people's life, whose objectives are to know and perpetuate The Original Law and the World's Heart, since the *younger brothers* (other indigenous and western people) cannot do so. Likewise, the name of Gonawindúa Tayrona synthesizes vital spiritual concepts within Kogui, Wiwa, and Arhuaco thinking. Gonawindúa is the name of the highest mountain in the SNSM in the Kogui, Wiwa, and Arhuaco languages. Its name is a metaphor for life, gestation, and fertility. It symbolizes the universe of the indigenous peoples of the SNSM (The World's Heart). Tayrona is the term in Spanish for Teyuna, which is related to a sacred space, The Lost City[9], and to an entity, Teyuna, the father of order and the good flow. In short, the OGT symbolizes the possibility of keeping The World's Heart (the SNSM) in harmony through the maintenance of traditional ancestral order.

On the material plane, the organization has established relations with the outer world through the *new leaders*[10] who have learned The Original Law, but also know how to interact with the outer world by knowing Spanish, national laws, and the nonindigenous institutional processes that affect the peoples of the SNSM and their territories. In short, these leaders, appointed by the *Mamas,* interpret the indigenous spiritual goals and exercise leadership in representing indigenous peoples to governmental institutions and recovering ancestral territory and traditions.

The OGT defined its general political purposes as the promotion of indigenous people's rights with respect to political autonomy and the recovery of ancestral territory and sacred places, such as the case of the Lost City.[11] These claims resulted from the historical territorial losses the indigenous peoples of the SNSM have suffered. These include the conquest in the 16th century; the influx of colonization at the end of the 19th century and beginning of the 20th century; evangelization processes; and the marijuana and cocaine boom in the 1970s and 1980s respectively and, more recently, the presence of paramilitary and guerilla forces and the national army.

The *new leaders'* role stands out in the process of claiming an ethnic identity. Leaders such as Ramón Gil, Adalberto Villafañe, Basilio Agustín Coronado, Margarita Villafañe, Cayetano Torres, Danilo Villafañe and Arregocés Conchacala, among others, have created an indigenous presence in local, national and international arenas through political proposals and territorial demands that promote respect for indigenous identity and environmental knowledge. At the same time, the OGT commenced a process of organizing local, regional, and national dialogue promoting indigenous issues, thereby allowing it to position leaders of the SNSM as legitimate representatives within the state.

In its 15 years of existence, this organization has demonstrated several approaches that express its negotiating flexibility and capacity from the local to transnational levels. As an organization, the OGT has undergone several adjustments throughout its history. It initially started as an occasional ad hoc authority in the city of Santa Marta and later became a broader institutionalized presence with formal councils and paid consultants with offices in Santa Marta. In the beginning, it focused on territorial recovery and greater participation in governmental institutions and programs. Later the OGT increasingly focused on establishing relationships with the CIT, as well as creating alliances with international organizations and other indigenous groups from different parts of the world.

Within the national context, agreements with state institutions allowed the OGT to develop working relations with institutional and national programs such as the Colombian Institute of Family Well-being (Instituto Colombiano de Bienestar Familiar), the National Rehabilitation Plan, National Learning Service, the Technological University of Magdalena (Universidad Tecnológica del Magdalena), and the Regional Corporation of Magdalena, among others. Within the international context, it initiated a relation with *Ricerca e Cooperazione,* an Italian NGO, and Alan Ereira, a British Filmmaker, who made the film: *From the Heart of the World: The Elder Brothers' Warning*, a BBC documentary (1991). Later he formed the Tairona Heritage Trust organization (see http://www.lamp.ac.uk/tairona). Likewise, Eric Julien, a French researcher, formed a foundation that seeks to purchase lands for the Kogui people.

Ricerca e Cooperazione had a very significant role in the OGT's consolidation. It supported indigenous communities in the northern slope by promoting indigenous culture and authorities, agricultural production for their own consumption, indigenous governmental political consolidation, and the coordination of actions among different institutions within the zone. The coalition with this NGO subsequently facilitated the creation of the "Gonawindúa Project" in 1992.

In 1986, the national environmental NGO called Pro Sierra Nevada de Santa Marta was founded (Fundación Pro Sierra Nevada de Santa Marta-FPSNSM). Its objective is "to promote and facilitate participation and agreement of settlers, communities, organizations and institutions of the SNSM and its influence area regarding sustainable development defined as the 'harmonization of human well being, cultural diversity and vitality of the ecosystems for present and future generations'" (www.prosierra.org). The OGT also established relations with this organization.

Within the national context, and as bound by the National Constitution of 1991, the OGT established alliances with governmental institutions such as the Indigenous Affairs office, Regional Corporations (Magdalena-Corpomag, Guajira-Corpoguajira), Incora, the national program of Rehabilitation (Plan National of Rehabilitation-PNR), and the regional governments of Magdalena and Guajira, among others, to develop the "Ecosierra Plan" and inter-institutional cooperation. The resulting inter-institutional agreements proposed solutions to the indigenous peoples' problems related to their legal territories (*resguardos)* and encouraged land reform in order to enlarge indigenous territories. They also proposed solutions to the problems of deforestation of the SNSM's lowlands and the consequent water shortages and other forms of environmental degradation in indigenous territories.

In 1992, shortly after the CPC-91 and the commemoration of the 500th anniversary of America's conquest, the OGT's members expressed their opinions of this event through a letter in which they highlighted the historical role of the SNSM's indigenous peoples in caring for the Heart of the World and maintaining the universe's balance. Likewise, they made a plea for political autonomy and the maintenance of their laws and customs, as did other indigenous peoples throughout the world.

During 1993 and 1994, the OGT continued its process of consolidation and maintenance of inter-institutional agreements with the support of *Ricerca e Cooperazione* and the national office of indigenous affairs. However, the OGT's political leadership underwent an internal crisis due to the OGT's political leaders' lack of authority from the perspective of certain indigenous interests. This process generated an internal division and the creation of a splinter organization, Avimolkueise, managed by an ex-member of the OGT. This organization disappeared later because all four of the indigenous peoples' organizations involved continued working together despite their differing priorities.

The OGT's territorial policies had their greatest impact in recovering sea access in 1994[12] when Cesar Gaviria expanded the Kogui-Malayo *resguardo* by presidential decree. This access to the sea reinstated an important element of indigenous rituals that serves as a demonstration of their symbolic control over their territories. Accordingly, the indigenous peoples of the SNSM designed their territorial policies to include the sacred and ancestral lands demarcated by The Black Line (La Línea Negra)[13] and the establishment of 39 sacred points in Presidential Resolution No. 835 of 1993.

Although the OGT had a series of important effects upon national and international arenas, it also had internal conflicts and moments of weakness because of problems with representation and leadership in its governors' council. Nonetheless, the OGT has established a series of relationships with several institutions and has positioned itself effectively within complex political negotiations (see tables 3.1 and 3.2).

One of the most important issues related to the OGT's policies[14] is its position on the Sustainable Development Plan (Plan de Desarrollo Sostenible-PDS). This plan proposes the following fundamental concepts to guide the national environmental policies pertaining to the "development" of the SNSM: (1) organization, coordination and adaptation of various public and private institutions, authorities and sectors to guarantee each entity influence over specific territorial settings; (2) respect for indigenous groups; (3) management of biodiversity according to the concepts of interculturality and integrality; (4) territorial and environmental ordering of the SNSM as a guiding principle of coordination and participation (identifying three zones: Kogui-Wiwa and Arhuaco indigenous *resguardos;* the National Natural Park SNSM that overlaps with indigenous peoples' territories and the remaining biodiversity zone where the MMA, the CARs, indigenous authorities, municipal and department authorities all intervene) that includes the zone of anticipated *resguardo* enlargement and the PNN buffer zone; and (5) the valoration and consolidation of existing coordinating authorities such as the Regional Environmental Council of the SNSM and the Directive Committee of the PDS.

This document highlights the indigenous peoples' role in a significant manner. On the one hand, it recognizes traditional indigenous cultures' ancestral knowledge as the best option and guarantee for the SNSM'S territorial and environmental management and, on the other, it states three goals as desirable for the future of the SNSM: (1) in the environmental dimension, the need to restore degraded ecosystems and to reestablish the massif's hydrological function and guarantee its water resources; (2) in the cultural dimension, to strengthen indigenous cultural identity and guarantee its autonomy in territorial management, and (3) in the territorial dimension, to have the traditional indigenous practices of natural-resource management take priority over other institutional forms of management.

According to the PDS, the principles that guide the proposal are respect, cultural biodiversity, equality, and participation. Based on the identification of the environmental problematic, and as a product of this process, the PDS has defined several action plans with specific activities aimed at overcoming the SNSM's environmental problems. These plans include:

Table 3.1. Principal Social Actors in the SNSM, Colombia

SOCIAL ACTORS	DESCRIPTIONS OF THESE ACTORS
Indigenous Peoples: Kogui, Arhuaco, Wiwa, and Kankuamo	Small settlements through lineages. The four peoples number more that 35,000
Local Leaders or Authorities	*Mamas* are spiritual leaders who also hold political power
Political Leaders	New Leaders, who enter into negotiations with governmental institutions
Indigenous Researchers	Professionals that lead local research processes and most of the time perform as public employees
Grassroots Organizations	Wiwa Yugumaiun Bunkwanarrwa Tayrona-OWYBT Organization, Kankuamo Indigenous Organization-OIK, Indigenous Confederation Tayrona-CIT and Gonawindúa Tayrona Organization-OGT and the Cabildos Territorial Council-CTC
Researchers	Anthropologists and biologists
National Environmental NGO	Pro-Sierra Nevada de Santa Marta Foundation
International NGOs	*Ricerca e Cooperazione,* The Nature Conservancy
Governmental Institutions	Ministry of Environment, Colombian Institute of Anthropology and History-ICANH, Indigenous Affairs, National Planning, government, mayors' offices, etc.
Multilateral Organizations	UNESCO, WB, GEF, European Union
Economic Sectors	Eco-tourism Programs Unlawful cultivators (Coca and marijuana)
Armed Actors	Paramilitaries, Guerrillas and National Armed forces

Table 3.2. Local, Regional, National and Global Importance and Recognition of the Indigenous Peoples' Territories in the SNSM

LOCAL, NATIONAL AND GLOBAL INTERESTS	RECOGNITION OF INDIGENOUS PEOPLES' TERRITORIES
Social Actors	Indigenous peoples, local leaders or spiritual authorities, native researchers, grassroots organizations, environmental national NGO, national environmental promoters, international NGOs, governmental institutions, multilateral organizations, economic sectors, armed forces, unlawful cultivators, and researchers.
Ecological Acknowledgement	Since the 60's the ecology-native relationship has been present in the SNSM. The Kogui were acknowledged by Bios as the most ecologically-conscious people worldwide in 1999.
Importance of the Region Environmentally	The SNSM is considered a site of great continental endemism due to its isolating characteristics and its geological formation on the Colombian Caribbean coast. Likewise, in current national environmental policies, it is considered a strategic eco-region.
Environmental and Cultural Policies of Indigenous Peoples of the SNSM	Kogui, Arhuaco, Kankuamo and Wiwa peoples numbering over 35,000 who live in the SNSM. Indigenous peoples of the SNSM have led many political struggles opposing development, education processes, and programs from governmental organizations. Indigenous peoples consider the SNSM as a sacred place, which has an ecological significance for the *Mamas*. The Sierra's indigenous peoples have always called themselves "elder brothers" and consider that they have the responsibility to keep the universe in balance through their practices.
Organizational and Political Processes	The Consejo Territorial de Cabildos and the four organizations of each indigenous people are organizations with local and regional political power even when these have been independent from a national indigenous movement. Indigenous peoples of the SNSM were important actors in marches and struggles of the national indigenous movement in the 70s.

(*continued*)

Table 3.2. (*continued*)

LOCAL, NATIONAL AND GLOBAL INTERESTS	RECOGNITION OF INDIGENOUS PEOPLES' TERRITORIES
	Indigenous peoples' political power lies in their political leaders as well as in their spiritual leaders and traditional authorities.
Superimposed Legal Forms	National Natural Park Sierra Nevada de Santa Marta (1977). Biosphere Reservation according to Unesco (1979). *Resguardos* (1980, 1983 and 1994). Archeological Park (Parque Nacional Arqueológico Teyuna-Buritaca-200) Lost City (National Monument 1995). Proposal of Humanity's Heritage as a Sacred Place.
Importance to the State	This region is of historical and archeological importance for the nation. Likewise, this zone has national political implications since it has a number of different economic and political forces: unlawful cultivation, paramilitary, guerrilla, and military. It also has ecological importance (see above).

ecosystem conservation, strengthening of indigenous cultural identity, peasant sector stabilization, strengthening of fundamental rights, and administrative modernization. Each of the plans has an educational component that addresses sustainable development and proposes programs and subprograms to meet a specific area's environmental and economic goals.

An important aspect within the PDS is the recognition of indigenous peoples' *resguardos* as a means of strengthening traditional indigenous environmental management because (the planners believe) the settlers and the peasants have contributed most to the SNSM's environmental devastation. The PDS also supports the view that indigenous peoples are most able to maintain or restore ecological balance. According to a local newspaper, the SNSM "has deteriorated due to external pressures from the acculturation process and invasion of *resguardos* with the consequent displacement of populations toward higher lands and introduction of foreign modes of cultivation, etc." (*El Informador*, June 29, 1993).

For these reasons, the indigenous peoples consider the PDS an important means to recover their lands as well as a means to conserve the Sierra. This is where the territorial ordering policy comes into action and where the constitution of Indigenous Territorial Entities gains recognition as an "option that will favor an adequate use of the Sierra's territory" (*El Informador,* August 23, 1999).

As part of the PDS, two projects were initiated: the Learning and the Innovation Project for the Sustainable Development or PAIDS (or LIL[15] in English) and the project of the Global Environmental Facility or GEF which has governmental and financial support from the WB.[16] However, because both projects are interrelated, they are subject to the WB's operating policy (4:20), which implies that the WB's employees define and verify the requirement compliance, not indigenous peoples.

THE CABILDO TERRITORIAL COUNCIL (EL CONSEJO TERRITORIAL DE CABILDOS-CTC)

Many indigenous peoples of the four organizations believed that the PDS does not express the indigenous vision, even though it has some formal aspects that seem to be pro-indigenous. Therefore, in 1999, they made a public declaration of the need for the four indigenous peoples of the SNSM to create an organization able to represent their values and interact more directly with the state and national society (Declaración conjunta de las cuatro organizaciones indígenas de la Sierra Nevada de Santa Marta para la interlocución con el Estado y la sociedad nacional).[17] The first step toward building this new organization and political process resulted in the consolidation of the four groups in opposition to the PDS.

> Two years ago, the Nation assumed as a national policy, a document well known in the region as the Sustainable Development Project of the Sierra Nevada de Santa Marta (PDS). This document was prepared from several workshops, where problem and solutions of the several sectors that live in the Sierra were set out. Even though the indigenous peoples participated in these workshops, this participation was not representative, since only a part of the communities that were not legal representatives of our concerns as indigenous people were present. For this reason, when our traditional authorities learned about the PDS, they noted that traditional thinking was not expressed completely as per the historical, political and social reasons we have, and they gave the

> order to our organizations to commence a broad consultation process that would allow necessary adjustments in order to restore the role of the indigenous people in the future of what western society currently calls the ecoregion. (CTC 1999)

In this statement the indigenous peoples of the SNSM also took a step toward instituting new decision-making processes legitimized by their own authorities, the *Mamas*. The *Mamas* were thus instrumental in raising the issue of indigenous participation and representation in the PDS meetings and programs. They also invoked identity and "authentic" indigenous thinking as necessary elements in any effort to generate alternatives to western plans for development. A basic assumption of the CTC's founding was the need to emphasize the environment as an integral part of indigenous peoples' cultural identity. Although the OGT recognized a special relation between indigenous peoples and nature, the joint statement clearly linked the issue of indigenous cultural identity to specific environmental issues involving the political autonomy they need to control their territories and natural resources.

This moment initiated a new kind of political negotiation between the Kogui, Wiwa, Arhuaco and Kankuamo peoples and national and transnational nongovernmental and governmental organizations. This new dialogue has sought to legitimize the political sovereignty or autonomy of the four indigenous peoples of the SNSM within their traditional territories (not just their existing *resguardos*), generating a series of implications that have prompted the development and consolidation of the cultural and environmental policies of the four peoples.

This statement marks the rise of the CTC as an organizational structure that enables the four indigenous peoples to coordinate and unify their demands for territories and rights of self-determination and autonomy according to their own cultural practices. As a result, it places indigenous peoples of the SNSM in the position of "nations" that can enter into a direct negotiation with the state with the sovereign power needed to make decisions regarding policies and activities within their territories. This power, however, creates the potential for conflict with other governmental development and investment plans that claim power in the SNSM, such as the Operating Plan of the European Union, the administrative program of the National Natural Parks System from the Ministry of Environment, and the investment and management projects of corporations and mayor's offices.

The CTC has pursued its goals by forming a permanent team to discuss indigenous political and environmental strategies and their conceptual definition. A person from each organization (that belongs to the CTC) forms this team: three external professionals acting as consultants (two anthropologists and one lawyer), the political leader of the OGT (Arregocés Conchala), and Danilo Villafañe and Cayetano Torres as indigenous consultants. However, the continued existence and efficacy of the team depends on the organization's financing, its political situation, and the conflicts between the paramilitary and guerrillas on its territories that have slowed and complicated its efforts.

The OGT's leadership has responded to the formation of the CTC by giving the Kogui people a dominant role in its activities. This has increased its legitimacy as a representative of indigenous interests insofar as the Kogui are viewed as the "most traditional" of the indigenous peoples and thus have more political authority in the national and international imaginary. This is reflected at the conceptual level in the documents which express the Kogui's philosophical ideas as paradigms to be followed by all the peoples of the SNSM. Consequently, the presence and leadership of the Kogui give the OGT the power to grant legitimacy to the actions and policies of governmental institutions, the NGO and international organizations that affect their territories.

The CTC's team formulates its policy from the ideas and interests expressed in each community, but its leadership is aware that internal consultation processes at the organizational level may generate policies inconsistent with input from the local levels, which in turn may create issues of authority and representation due to the contradictions between new leaders and local authorities and the cultural transformation of indigenous values that results from these consultative processes. Therefore, they continually work to produce the consensus needed to legitimize the leadership's authority.

The CTC also focuses on generating general policies designed to interact with nonindigenous institutions. These policies start from the basic consensus of the four groups founded upon the *Mamas* laws. The assumption is that all negotiations will start from the *Mamas*' principles when interacting with institutions or programs in which the relationships are mediated by money or other nonindigenous measures of value (commercial agricultural production, efficient resource extraction, for example).

CTC's Environmental and Cultural Politics

The CTC proposals have arisen as an alternative to external notions of macro-regions, eco-regions, rural zones and sustainable development projects in which indigenous peoples' notions of territory as a sacred space that integrates the human and nonhuman have been lost. According to the indigenous peoples' perception, notions of territory as integral should take priority over the external actors' views of territory. These (NGOs, state institutions, and private companies) conceive the SNSM as a resource, an unused empty space, the common heritage of the nation or humanity, or a number of private spaces for each social actor. Lastly, the nonindigenous perspectives assign each level of the physical space in the following manner to a certain social group: the high part of the SNSM belongs to indigenous peoples, the middle part belongs to peasants, and the low part belongs to urban populations. This divides the indigenous peoples' notion of territoriality and presents the sacred sites as isolated points rather than integral parts of the landscape. These concepts ignore the historical relation between territories and culture that indigenous peoples claim and use as the basis for their identity.

Therefore the CTC argues that there will not be any spatial or political division in the relationships of indigenous cultures, since the SNSM cannot be seen in a divided manner, but is a unit that is sustained in the Original Law that unifies the four indigenous groups into one. This integral and autonomous notion of territory depends on the coordination of the CTC and the *Mamas* for its political power.

The CTC initially used its political power to criticize the PDS' project for promoting governmental policies of sustainable development that had not been formulated with local participation and whose legitimacy therefore existed only at the national level. One of the main critiques has been that the PDS presents the SNSM's problematic and its different actors without the historical context that explains the process and the presence of these actors. Therefore, as a counterproposal, the CTC presents the territory in a historical manner that addresses the different intervention processes that account for the presence of nonindigenous social and cultural groups and that recognizes the autonomy that indigenous groups have had historically. In this manner, the CTC's process of political organization has allowed the clarification of its objectives and a focus on the generation of policies and new strategies to deal with the state based on indigenous values that differ from those of the PDS.

This political context suggests the following scenario: a tacit agreement on indigenous peoples' legitimacy as nature's protectors supported by the MMA, the National Natural Parks Unit and the governmental environmental corporations (based on indigenous participation policies and nonindigenous sustainable development policies); and the explicit recognition of the right and effective political power of the indigenous peoples to control their territory (sustained in the CPC-91 and the ILO-169).

Indigenous peoples, through the CTC, have expressed their policies in a set of draft documents that trace the open-ended negotiation process resulting from the fact that each of the four indigenous groups have their own organizational, social and political dynamics, which effectively make agreements on each document an ongoing process. The indigenous peoples see these draft documents as "fundamental instruments of indigenous peoples' efforts to define relations and coordinate activities with the nation concerning traditional territory or, better yet, the Sierra Nevada" (CTC 2002).

These documents consistently employ indigenous cultural principles and concepts as the framework for negotiations and meetings with the nation regarding the SNSM's environmental management. In other words, this indigenous policy framework assumes the indigenous government's exercise of control over traditional territory. In addition, the indigenous character of these documents allows the four indigenous groups to reach agreements in a more explicit and rapid manner among themselves.

These Spanish-language drafts are the "official proposals" that support all the negotiating processes among the four indigenous organizations and governmental institutions, but they remain flexible and modifiable. This is so because the four indigenous cultures speak different languages and they are not familiar with written Spanish (even though they may speak Spanish among themselves). Therefore, each draft has to be translated and interpreted in each language and then discussed in Spanish to create a general discourse. These documents thus require a permanent negotiation among the four groups who may then agree upon policies and actions that serve their interests.

Indigenous peoples' express the environmental and cultural principles of their politics in their declarations or "texts." They articulate these principles as an idea of integrity that promotes environmental conservation as a "reciprocal harmony and order among natural elements and human beings" (CTC 2002). The indigenous position in the negotiating process with the nation has taken shape as a political proposal with three main bases: territory, traditional authority, and a traditional way of life that assumes political self-determination. The strengthening and consolidation of these three bases, through specific actions are the pillars that they have negotiated with the nation as the basis for an environmental plan for the SNSM. It is within this framework that indigenous peoples believe that they may promote the conservation plans, territorial integrity, and cultural permanence that will fulfill their life plans.

In other words, indigenous peoples of the SNSM have established their proposals by synthesizing their contemporary governmental criteria and processes with traditional indigenous conceptions of rights and territory (CTC 2002).[18] The elements of this synthesis may be specified as follows:

- The Original Law (Ley *Sé*): in which "the spiritual and the material is condensed and serves as a guarantee for territorial permanence. It is also a balance and a base for the territorial control and management that was awarded to us. It is where the cultural identity of our people resides." Through this Law indigenous peoples became the territory's guardians, and they therefore feel a great responsibility to serve in that capacity.
- *Eswama:* This refers to the "space chosen by the mother to make great decisions" that coordinates the indigenous peoples' concerns and commitments. *Eswama* is the collective principle for the use, management and care of the territory, and it is where the Original Law is interpreted, translated, and experienced.
- *Mamas:* These are men with greater knowledge; "they analyze, interpret, and transmit the Original Law's instructions."
- Indigenous Organizations: The *Mamas* delegate the CTC to act as a competent authority in relation to state affairs and other actors outside of their culture. Through this agency, they seek to "guarantee coordination and harmony of the external acts affecting traditional ones."

The categories on which the legitimacy of ancestral ownership depends may be specified in the following terms:

- History: One that is "based on the continuity of harmonic and balanced responsibilities, where there is not any ignorance or domination."
- Interdependence: This means the relationship between indigenous territory and culture.
- The Sacred: There are "practices based on the ancestral law where thinking and nature are joined, and this thinking is the permanence of an established order and necessary balance."
- Life: "Within the territory every thing is a bearer of life, or content of the *Law Sé,* where the existence of everything is found."
- Order: "The territory is a system integrated by sites, knowledge, and practices determined in the *Law Sé* and supported by knowledge, laws and their management."

- Responsibility and competence: This the part of the *Law Sé* constituted by the *Mamas, Makú* and elders, which exercises authority, and which the indigenous organizations delegate to the CTC for external management.
- Ancestry and tradition: This territorial category determines land ownership.

They also ground their rights in the recognition that the state has given to their plan for the SNSM's management as the most adequate to conserve its unique ecosystem.[19] They reinforce this argument with the statement that the National Natural Parks are on traditional territories precisely because of the management that they have given to their territories. They also state that the recognition of the Biosphere Reservation on their territory is a result of their management. Consequently, they believe that their relationship with the environment is the best way to maintain its ecological balance because the philosophical principles that guide their activities have in fact produced that result. According to Ramón Gil, a Wiwa:

> We are the ones here to be in charge of taking care of nature: the birds, animals and minerals of the Sierra Nevada that for us is called *Ima Ugumakcanna,* that which is of the mountain. There everything was born. There was the birth of all wisdom and laws.

As mentioned before, the cultural and environmental policy of indigenous peoples of the SNSM is based mainly on strategies to consolidate territoriality, authority and traditional life (self-determination), which underscores how their concrete environmental practices arise in a "natural" manner from the synthesis of these cultural strategies and the environment: "Territorial consolidation of the indigenous people is a natural conservation and protection strategy, and is grounds for its cultural strengthening based on the Black Line" (CTC 2002). This territoriality marks an ancestral relationship that leads indigenous peoples to propose an integral management (environmental and cultural) for all the territory in order to assure both its environmental and cultural conservation. Thus, territorial consolidation is a natural conservation and protection strategy, and it is the basis for the cultural strengthening of the Black Line within which cultural identity becomes manifest through "education and the health of nature that promote permanent standards for the balancing and ordering of human beings and nature" (CTC 2002).

An important element that has to be highlighted in addition to those noted above is the right claimed by indigenous peoples to "actively participate in a decisive manner in each and every activity and project that affects the integrity of our traditional territory marked out by the Black Line and the sacred sites that define it" (CTC 2002). Consequently, rights acknowledged by the ILO-169, Law 21 of 1991 and the CPC-91 are consistent with indigenous peoples' traditional understanding of territorial management, and indigenous peoples assert that their traditions dictate that they cannot renounce their right (responsibility) to make decisions regarding their territory.

From this perspective, the territory is seen and felt as an experience of the sacred in daily life, of knowing the laws of relations with other beings who may or may not be human, and of the management of relations that involve them. Each part is interrelated to the other, and they cannot be separated or sold. This territorial notion links and coordinates physical spaces in both symbolic and daily life.

Unlike many indigenous peoples, those of the SNSM distinguish themselves by their use of formal political organizations and actions. They prefer to establish their positions in public declarations and political arenas rather than public mobilizations or protests. By doing so, they project an image of themselves in the national and international media as ecological indigenous peoples whose great care for nature is based on their ancestral wisdom from which the "younger brothers" should learn. Their strategies include seeking relations and support from NGOs, environmental organizations, human rights organizations, anthropologists and public figures who can help publicize their concerns about their cultures and territory. Indigenous peoples have a clear awareness that their cultural identity has been affected by the importance of the SNSM's status as one of humanity's heritage sites due to its biological and cultural diversity. This linkage has become an important political strategy for publicizing their status as autonomous indigenous peoples. In other words, they know that their representation as "ecological indigenous peoples" gives them a particular position of power and authority within transnational arenas.

CTC's Politics and Negotiating Dynamics

The CTC has used the Development Plan's Directive Committee and the Environmental Regional Council to promote their proposals related to environmental and political issues in their territories. During the years 1999–2003 the CTC has transformed negotiation and participation processes by gaining the power to make decisions and proposals that influence or determine those of other indigenous and nonindigenous governmental and nongovernmental institutions.

This power and authority arises in large part from the following sources:

Governability and self-determination: The CTC's political negotiations are based on the right of indigenous people to have autonomy and self-determination over their territories and their resources. The CTC positions itself as a legitimate authority without calling for a separation from the nation. Instead it calls for coordination and the construction of a multi-cultural and multi-ethnic nation that will respond to a diversity of governmental institutions and legislative processes.

The legitimacy of the indigenous peoples' representatives due to support from traditional spiritual authorities: This is a significant element since the visible political head, Arregocés Conchacala (Kogui), has the prestige and legitimacy of being related to the *Mamas,* which gives him legitimacy both internally and externally. In this situation, the Kogui people and traditional indigenous people's authorities gain recognition of their leadership and authority from national and international institutions due to their ancient wisdom and to the management of their environment. This is one of the elements with the greatest credibility for the outer world because it coincides with the conceptions and expectations that the West has of an *ecological native.* This is reinforced explicitly through the use of their dress, called a "manta" (Kogui and Arhuaco), and their own language through which they represent indigenous tradition and authenticity to the western public. However, their recognition arises not only from fulfilling western ideals, but also from the fact that indigenous peoples are legitimate agents who know their territories better than nonindigenous peoples due to their historical and cultural relationship with it. In short, they need to publicly and visibly link their ancestral indigenous knowledge to the environment to gain power in nonindigenous contexts.

The unity of the four indigenous groups: The unity of the four groups promotes a common agenda that allows the establishment of a sole representative, thereby preventing the division of indigenous peoples' authority. The generation of a coherent and unified discourse concerning the integrity of territory, nature and culture gives a philosophical basis to the arguments for indigenous peoples' integrity within the Black Line and this, in turn, has legitimized ancestral knowledge as the basis for the integration of external actions and interventions.

The direct and equal coordination with national authorities: The indigenous people have highlighted the need to have coordination with national authorities so that a horizontal dialogue from authority to authority is established.

The continuity and maintenance of their regulations for interaction with nonindigenous institutions: The consistent use of an indigenous framework for evaluating the intentions and actions of external actors.

The rejection of being mediated by NGO's or external consultants: The indigenous claim autonomy in their actions and control of their policies. One of the beliefs that has helped most to empower indigenous peoples is their certainty that external actors cannot speak for them. Therefore, they resist nonindigenous manipulations of their internal decision-making processes.

Other indigenous organizations have been more open to nonindigenous participants. Currently the OGT employs two anthropologists and one lawyer as advisors; and the CIT has an anthropologist and a legal adviser as well. The CIT has developed workshops and seminars where the most important *indigenistas* within the national context have been advisors. The OWYBT has also consulted an anthropologist, while the OIK has worked with two anthropologists.

FINAL REMARKS: THE POLITICS OF SPACE AND DIFFERENCE IN THE SNSM

The CTC's political strategies may be characterized as what Leff (2002) calls "the politics of place, space and time" through which indigenous peoples' communities have promoted the opening of more democratic participatory arenas. For example, the CTC has opened participatory spaces in the PDS Directive Committee, in the Regional Environmental Council, and in institutional meetings attended by nonindigenous representatives, consultants, spokesmen and the voters from the four communities and their representatives. These interactions between the CTC and the state institutions have established agreements and differences and have fostered new strategies affecting territorial, cultural and environmental issues that have repositioned indigenous peoples of the SNSM as agents able to propose alternative territorial and environmental relationships.

Indigenous policies find common ground in the territorial notion that links and coordinates the identity of the four indigenous peoples and through their shared responsibility for the SNSM's environmental and cultural conservation. Claiming their traditional right as environmental authorities has thus enabled them to exercise control over their territories and "resources," and the exercise of this right has generated proposals and alternatives to sustainable development. According to Leff (2002), such indigenous alternatives construct *cultural territories* that allow culture to serve as the coordinating political principle for environmental management and conservation strategies.

Such a cultural territory is evident in the CTC proposal to develop a notion of territory based on The Original Law. This notion consists of a reading traditional meaning of the sacred sites and their signs to determine how to act in the present.

> In this way, indigenous people are asserting their cultural rights to recover control over their territory as an ecological, productive and cultural space and to regain possession of a heritage of natural resources and cultural meanings. The environmental rationale is being internationalized by new social actors, expressing it as a political demand that guides new principles to value the environment and to reappropriate nature, taking root in new territories and new identities (Leff, 2002).

The CTC's proposals thus promote an environmental and territorial knowledge whose cultural values and practical expressions have permeated the proposals and affected the activities of a variety of nonindigenous actors (mayor's office, corporations, NGO, state institutions). The CTC's proposals have, for example, sponsored cooperation with the National Natural Parks Office and the National Planning Office in their efforts to rethink and redesign programs proposed for indigenous territories. In addition, the Foundation Pro Sierra Nevada de Santa Marta has stopped projects that violate its principles. For example, the CTC's policies stopped the "The Unique Operating Project" of the European Union and the GEF Project by pointing out the lack of indigenous participation in their planning. The Word Bank, which administers the GEF program, had to intervene and recognize that, according to its own operating policies for indigenous peoples called the 4:20, the beginning of the project lacked indigenous peoples' participation.

Although the CTC has general guidelines, the practices and actions proposed differ according to each indigenous group and community. For example, in the Koguian situation, communities are organized as independent units (micro-powers). Therefore the programs and projects developed for their territories differ from the Arhuaco peoples' projects. In this way, the indigenous peoples are influencing and changing the global view of sustainable development by setting out environmental strategies according to local values and conditions as they understand them.

In short, the emergence of the environment as a global concern and political arena has given rise to new social, political and cultural processes that have yet to be fully analyzed in terms of how they affect the relations of the indigenous peoples and territories of the SNSM to the various nonindigenous participants in that globalization. These processes involve negotiations regarding access to genetic resources, collective territorial and intellectual property rights and global policies of sustainable development; and they transform the lives and cultures of indigenous peoples insofar as they require a new indigenous politics situated at the transnational level. The following chapter focuses on the emergence of this global and national environmental awareness and how indigenous peoples respond to it and are affected by it.

Chapter Four

Thinking Green: Global Eco-Governmentality and Its Effects in Colombia and the Sierra Nevada de Santa Marta

INTRODUCTION

Western representations of indigenous peoples have been characterized historically with an emphasis on their preindustrial (underdeveloped) ecological practices and otherness with respect to western economic progress and cultural "normalcy." Only recently have western discourses represented indigenous peoples as equals (if not superiors) due to their status as valued allies and participants in the cause of environmentalism. There are multiple reasons for the West's newfound interest in this coalition. Much of this interest is due to the efforts of indigenous peoples themselves: indigenous political strategies that emphasize the environment; indigenous discourses that reposition indigenous representations in the western imaginary; the practical interventions of indigenous practices within nonindigenous environmental and developmental discourses. For its part, the West has contributed changes in its conceptualizations of indigenous peoples and nature as elements of this growing interest: an awareness of the necessity of constructing a new society due to the crisis of industrial economic development; the epistemological shift in natural and social sciences concerning notions of nature; the desire to profit from the territories and resources of indigenous peoples within its commercial circuits, and a growing awareness that indigenous people indeed often have better ideas about environmental management, among others. However, the principal reason for this coalition is the growing awareness in the West of the global environmental crisis that it has caused and the consequent emergence of efforts to formulate alternatives to old ways of interacting with "nature." The result, from the western perspective, has been a steady movement toward a global eco-governmentality.

Environmental awareness appears in our daily lives. We (in the western world) are now accustomed to finding recycling signs everywhere that imply new relationships with the use of natural resources. Children discuss the problem of global warming daily in schools and households. Politicians and experts around the world discuss issues of ecological management and preservation. Governmental and non-governmental organizations consider environmental problems to be "global" and "universal" human concerns. Indigenous peoples who live in environments not so badly affected by our environmental problems appear to us to be the people who know how to deal with the environment.

Increased environmental awareness around the world attests to the evident deterioration of ecosystems and the depletion or extinction of natural resources that have transformed not only ecological processes but also cultural practices and the intricate network of meanings constructed around the idea of the environment. This problematic is also related to the political actions of the environmental movements, which have helped to transform the social and natural disciplines and change the relationships between society and nature.

However, all of this implies new forms of interdependence and global policies related to natural resources or eco-governmentality. Therefore, it is necessary to trace historically successive waves of environmentalism to understand the formation of the current discourses of eco-governmentality. I argue that we are facing an eco-governmentality in which new regulations related to property rights arising from western environmental discourses are presented as necessary to the defense of the planet and its biodiversity. This eco-governmentality is evident in Colombian national policies, and within this new eco-governmentality indigenous peoples, in general, and indigenous peoples of the SNSM, in particular, have become important actors because their territories and "natural resources" have become focal points of global environmental interest.

Therefore it is useful to make explicit the premises that are involved when discussing local and global eco-governmental discourses and activities. This includes recognition that this new discourse and mode of reasoning has had particularly marked effects on indigenous peoples and local communities who have assumed (or been assigned) roles as guardians of nature or *ecological natives*.

This chapter is a journey though the emergence and development of global environmental awareness that explores its different versions and conceptions, with special emphasis on the role of environmental movements and their main tendencies: the anthropocentric and the biocentric perspectives. It also explores how environmental discourse has become a global environmental strategy (eco-governmentality) that has affected Colombian environmental policies. Finally, the chapter addresses the development of specific environmental discourses and movements in Colombia and, in particular, the SNSM, and their interrelation with global environmental policies.

THE EMERGENCE AND DEVELOPMENT OF GLOBAL ENVIRONMENTAL AWARENESS

The emergence of environmental awareness has been dated in different ways. Some authors locate the origin of environmental thought in humanity's beginnings and point to "natural" responses in human feelings toward nature guided by religious, ethical or philosophical principles. These authors present environmental concern as a human response that transcends cultural differences and notions of time and space (Dobson 1989), and they note that these principles have gained importance recently because of the environmental crisis. Other scholars locate the beginning of environmental awareness in the 18th and 19th century as an answer to the problems that were generated by the process of industrialization (Luke 1997, Guha 2000). Finally, for most scholars, the environment is a new category that appears in public discourse during the 1960s and 1970s in relation to "environmental" movements and their political actions (Dobson 1989).

The Industrialization Process and Nature

The industrial revolution in England transformed the landscape during the 18th and 19th centuries through urbanization and industrialization. The industrialization process and the consequent consumption of natural resources caused new problems, such as the conservation of spaces for hunting, increases in the human population, dislocation of rural people, pollution, and the extinction of species. As a result, concern arose among intellectuals, scientists and some government officials regarding the protection and conservation of nature. Although they tried to influence the rulers of European and North American nations, the initial proposals of these "environmentalists" did not have wide support. Their main focus was the conservation of water and forests and the criticism of the environmental and social impact of modernization, particularly industrialization.

Guha (2000) notes that there have been two waves of environmentalism. In the first wave in the 18th and 19th centuries three tendencies arose which reemerged in the second wave of environmentalism in the 20th century: First, the moral critique of industrialization (*the return to the earth*) that called for a return to simpler forms of pre-modern life; second, scientific conservation based on the efficient use of resources; and third, the idea of wilderness based on the notion of the preservation of pristine or virginal nature.

The environmentalism of the 19th century generated responses from poets and writers who clamored for a more rural lifestyle. The works of British poets like William Wordsworth (1770–1850), John Clare (1793–1864), John Ruskin (1819–1900), William Morris (1834–96) and Edward Carpenter (1844–1929) exemplify the romantic critique of industrialization that proposed as an alternative a self-sufficient communal-agrarian life. Ruskin emphasized the consequences of pollution on humans while Carpenter internationalized the proposal of returning to the earth in his *Civilization: Its Cause and Its Cure*. This book is a fundamental document for environmental movements. Many of the premises of environmentalism were also articulated in the works of North American writers such as Walt Whitman, Ralph Waldo Emerson and Henry David Thoreau. Likewise, Octavia Hill's (1838–1912) proposals are very important to environmentalism because they link environmental protection with social reform. She carried out her beliefs in concrete environmental actions in poor urban areas and also organized public demonstrations against pollution. In the German romantic tradition environmentalism was linked to a nationalism in which the peasants, the forests and the nation were an organic whole. The German industrialization process thus generated responses from poets and intellectuals such as Rainer María Rilke who stressed the role of the peasants' agrarian life as an important basis of national identity (Guha 2000).

Faced with the decrease or destruction of natural resources, preservationists and conservationists in different western countries developed various strategies. At the end of the 19th century, some countries had established restrictions on the use of natural resources and witnessed the creation of natural parks. Local and national interests thus began to generate new ways to control and use nature under the logic of the principles of scientific conservation and management.

Scientific conservation can be traced to George Perkins Marsh's book *Man and Nature* (1864), considered a basic text for American environmental movements. It is also related to the works and ideas of Dietrich Brandis, a German researcher who directed the forest service in India. These two scientists shared the idea that the power of expert knowledge could repair environmental damage by promoting the sustainable use of forests, fisheries and wild animals. They were the pioneers of a scientific movement with global consequences, and their vision was possible because the centralization and conformation of the modern state allowed the intervention of expert knowledge of natural-resource management on a national scale and because "natural resources" had become the property of the modern state and valuable to the industrial interests that dominated it. These processes also allowed the consolidation of new disciplines and expert knowledge (that of hydrological engineers, resource conservationists, managers of wildlife, among others). These experts also managed the resources of the European colonies in Asia and Africa. The objective was to order "nature" through state controls and special rules formulated and administrated by state-sponsored experts in order to promote progress (Guha 2000).

The philosophy that has sustained scientific conservation promotes the idea that an environmental crisis is imminent unless current practices are replaced by more "rational" (nontraditional, nonideological) uses of resources. State-sponsored expert knowledge therefore often assumes the view that traditional local practices are inadequate and wasteful, which requires their transformation through expert knowledge and rational planning under the "management" of state experts. "However, scientific conservation was an ideology that was at once apocalyptic and redemptive. It did not hark back to an imagined past, but looked to reshape the present with the aid of reason and science" (Guha 2000:30). In this sense, it is the polar opposite of the romantic view of pre-industrial relations to nature.

In this vision was implicit the idea of a unique and sovereign state able to take a long-term view of managing the forests and the water according to technical regulations and proper rational use. In this perspective, conservation is an instrumental calculation: one of correct use, efficiency and development. This implies that peasants and local people are ignorant and that they are to be condemned for the unscientific management of their resources. It also implies that management is apolitical.

The Germans also promote studies of quantitative calculations that allowed for the estimation of sustainable crops. They coordinated the departments of forestry science in almost all of Europe, which also allowed more efficient resource extraction in the colonies. Likewise, in 1890, the German influence appeared in the American Forest Department when Gifford Pinchot assumed its direction in accordance with Dietrich Brandis' ideas.

In the colonies, scientific forestry management was based on the idea of "surveillance" that strengthened the state's control and often outlawed the rights and practices of the peasants and traditional communities. In fact, the forests under state control were wild spaces forbidden to community practices, which generated conflicts of vast proportions. In India, for example, the forest laws established in 1878 generated local protests aimed at recovering indigenous control over natural resources. Likewise, the political, environmental and cultural differences in other colonies interfered with the implementation of sustainable harvest programs based on expert western knowledge of forest science (Guha 2000).

At the same time, these policies implied new ways to administer time and space for humans that promoted access to natural resources based on scientific knowledge and new industrial methods of controlling and managing such resources. The power of the natural sciences provided the knowledge required for the implementation of these new industrial strategies for managing natural resources. Such resources were mapped, catalogued, analyzed and assigned specific locations in the objective space and time of science where they could then be valued and appropriated according to their potential uses in the modern project. Thus, the notion of property and ownership of nature became possible on a national and even international scale. This conception prompted both businesses and the state to seek proprietary rights over nature: the states reserved natural parks as recreational areas for nations, and businesses sought extraction rights over resources. Industry, however, was the major beneficiary: large regions became the properties or resources of international industries (corporations). However, scientific experts laid claim to some forests and fishing areas which they proposed to maintain in "natural states" through expert scientific management designed to propagate flora and fauna of value in the national or international market places.

The modern conceptions and methods of controlling nature established a specific relation of separation between humans and nonhumans that implies that nature is an object subject to human management. The self-interest of humanity, according to this view, dictates that the development of nature will occur according to the rational calculations and technical processes of science that will prevent the depletion of resources or environmental damage harmful to humans (enlightened self-interest). This reflects an anthropocentric vision in which nature is at hand for human consumption. It also expresses the industrialization of nature understood as raw material for development through a set of technical and instrumental processes. The modern idea thus located nature at the basis of its ideal of progress toward an improved future. Controlling nature was necessary to the industrialization and material standards of living that would allow people to become "truly modern" or civilized. In this sense, the scientific idea of nature and its management occurred as an ideological element of the discursive formation called "modernity".

The idea of the wilderness as a place reserved for the protection of a pristine nature is also a modern notion, but in this case one that results from nostalgia for the premodern based on a quasi-religious or romantic view of nature. This notion has its origins in the images of sacred places, such as caves or places in the forest that deities and spirits inhabit according to beliefs of non-Christian cultures.[1] It also has origins in the spaces for hunting in the feudal times that were dedicated to the pleasure and delight of the elite. In modern times, national parks became the expression of such a desire for wilderness. Yellowstone (1872), the first national park, was founded as a strategy to conserve what little native forest remained after the exploitation of the colonists and as a nature reserve for future generations. Among the American philosophers and activists who promoted the idea of wilderness were John Muir, who founded The Sierra Club in 1892, and promoted the philosophy of ethical preservation, and Aldo Leopold, who promoted the protection of wilderness outside of the parks.

The interest in protecting natural resources was also a response to the development plans of colonial powers. Conservation in the colonies strove to create management strategies coordinated with the industrial economy of the empires. In 1900 in London the first international meeting on wildlife in Africa took place (Convention for the Preservation of the Animals, Birds and Fish in Africa). This meeting ended with the formation of the "Society for the Preservation of the Fauna of the Empire" in 1903. These conservation strategies did not consider the local people's interests and were fundamentally long-term strategies driven by western values and interests.

> The first step was to moderate demand by specifying closed seasons when animals could not be shot, and using licenses, the possession of which alone allowed hunting. The second was to designate particular species as 'protected.' The third step was to designate specified territories as 'game reserves' meant exclusively for animals, where logging, mining and agriculture were prohibited or restricted. The final and most decisive step was the establishment of national parks, which gave sanctity to entire habitats, not merely to animal species dwelling within them. (Guha 2000:46)

This approach, in addition to the others mentioned above, constituted the first wave of environmentalism. However, these topics were not of genuinely international importance during the first half of the 20th century because of the historical, political and economic conditions at that time had not yet become a truly global crisis.

Environmentalism and Environmental Movements during the 1970s

During the 1960s and 1970s the depletion of species and natural resources increased and new industrial processes generated more pollution and hazardous wastes whose harm to the environment began to affect human life in more intense, widespread and disturbing ways. The beginning of the contemporary environmental movement can be marked with the publication of Rachel Carson's book *Silent Spring (*1962). As environmental movements emerged demanding new relationships between humans and nature, novel forms of environmentalism also emerged.

Some scholars such as Dowie (1995), Williams and Mathenny (1995) and Luke (1997) identify North America as the first country to initiate political actions aimed at the protection and conservation of "the environment." Luke (1999) claims that during the 1960s there was a pronounced conceptual shift in America from nature to environment. He notes how the concept of "environment" began to be used during the 1960s and how, from this perspective, the environment began to be considered as distinct from natural resources and as something that might be valuable in terms of its impact on the quality of human life rather than its impact on levels of consumption.

Carson's book criticized the specialization and fragmentation in the natural sciences' approach to environmental problems and articulated the implications of maintaining industrial priorities over environmental ones. This criticism subsequently developed in discussions on the limits of growth at the Club of Rome (1972) and the positions of scientists (Brejér, Huxley, Schumacher, among others) and philosophers (Arné Naess, Mumford, among others) who called for the protection of the environment. These critics brought about the second wave of environmentalism that manifested itself in the three tendencies identified by Guha (the moral critique of industrialization, scientific conservation and the preservation of the wilderness). In this second wave, philosophers, scientists and environmental movements' activists shared the idea that the industrial economic system and its development processes were causing the environmental crisis (Guha 2000).

In the 1970s the idea of wilderness was revived through the Neo-Malthusianism of Hardin and Ehrlinch who addressed how the human population affected the space of other species. This also concerned the followers of Arné Naess' ideas about the impact of humans on nature. In contrast, the tradition of scientific conservation strengthened the Club of Rome proposal that sustainable development should become the alternative to progressive economic development. This view would require scientists to develop alternative technologies and states' policies to place controls on production processes. As a result, the environmental movements' views entered the political context insofar as they questioned the existent relationships between science, society, industry and the environment.

Radical members of conservationist and preservationist organizations, as well as experts, technicians and members of the middle class also helped create the environmental movements that emerged at the end of the 1960s and beginning of 1970s.[2] Their discourses and actions focused on opposition to development projects (the construction of highways, oil fields, nuclear reactors and hydroelectric dams, among others), urban centers of great scale, and pollution.

These collective actions prompted scientific discussions about health, toxic waste and technical matters related to environmental problems. The values of these movements also manifested themselves practically through a combination of legal mechanisms (the imposition of sentences, fines, injunctions upon states and industries) and radical protests (from occupation and obstruction of development projects to violent confrontations).

As with the emergence of other social movements of the 1960s, the environmental movements grew in the countries of the north as another instance of the desire for more democratic participation in political arenas (civil rights and anti-war movements), and as a result environmental movements shared the types of strategies and actions used in other movements (marches, mass demonstrations). The first celebration of the "Earth Day" on April 22, 1970 was a mass celebration of the environment in the northern hemisphere. From these beginnings, environmental movements have grown considerably, and they now have the support of millions of people around the world who advocate new relationships between nature, government, the economy and society. However, radical environmentalists claim that the movement in general has been bureaucratized to the detriment of its initial purpose. Because of this, some activists defend more radical proposals for the protection of nature.

Since the 1970s, diverse environmental movements have been formed, from Chipko to Greenpeace, which have been opposed to scientific forestry and development programs, such as hydroelectric and industrial tree-farming projects of great scale. Similarly, local grassroots and ethnic social movements have joined the environmental cause. They offer their traditional environmental knowledge and practices as alternatives to the imposition of external environmental management and development programs. In some cases, they have consolidated themselves with national and international environmental movements and government agencies thereby generating global eco-bureaucracies.

The actions of environmental movements have generated new political processes and changes in the conceptualizations of nature through their promotion of environmental problems as a topic of public interest. Such environmental concerns have in turn encouraged scientific experts to address the environment, especially in the social and natural sciences, thus forming the possibility of new discourses and theoretical perspectives. In this sense, the environmental movements have been successful in questioning modern conceptions and relationships concerning nature, development and progress and making the conservation of the planet a basic topic of global politics. In other words, they present a political model that criticizes the global strategy of limitless growth and modern rationality's subordination and domination of nature. However, environmental movements cannot act alone, and the resolution of the conflicts surrounding the environmental crisis implies ecological and social programs that transcend legal actions, protests and studies. Therefore, their objectives have expanded to include governmental policies and programs aimed at changing development models and other activities related to the environment. Insofar as these policies and programs challenge the underlying assumptions of modernity, they encounter strong opposition from industrial and investment interests that benefit from the status quo. For these reasons, the process has been slow and, on occasion, not very effective, as is evident in the fact that international agreements such as the CBD have not been ratified by all states (Alario 1995).

As the new environmental movements began to confront entrenched economic and political regimes, they began to question the link between science, processes of industrial production and environmental degradation. This prompted a rethinking of the relationship between society and nature that required changes in the rules for producing and evaluating the worth of scientific truths and knowledge. Consequently, these changes also promoted further discussions about the legal bases for access to natural resources and their ethical use. Diverse social movements, grassroots organizations, women's groups, animal rights activists, among others, also began political actions against the processes of industrial development, contamination and extinction and thus opened yet more arenas for discussion of the environment.

These changes also led to an interest in indigenous perspectives on the natural world. As a result, indigenous peoples' political struggles to maintain their cultural practices and relationships among humans and nonhumans and against the capitalist patterns of consumption have also contributed new approaches to the environmental problem. In fact, academics such as Shiva (1994) and Escobar (1999) identify the practices and struggles of indigenous peoples to preserve and promote their views on human relations to nature as the true beginning of environmental awareness and discourse.

I have presented the discourses and actions of environmental movements in general terms. However, specific discourses on the environment differ among themselves greatly and employ diverse conceptions of nature and ecology as well as diverse theories drawn from the natural and social sciences, all of which implies that specific environmental movements are diverse in their values and goals when examined more closely.

CONCEPTIONS OF NATURE

Diverse notions of nature coexist in specific social settings and according to particular historical moments. Notions of nature inscribed in monist or dualist visions have been present in diverse times and spaces. However, certain notions of nature have risen to nearly complete hegemonic status in some historical moments. The modern conception of nature, with its emphasis on progress, industrial development and efficiency, has achieved such a status and has influenced thoroughly the scientific conception of nature and, of course, anthropology's "scientific" investigations of indigenous peoples (Ulloa 2001a, 2002a, 2002b).

Two concepts have generally dominated conceptions of nature in the West: monism and dualism. Monist conceptions of nature were basic to the Renaissance ideas in which nature and humanity were part of a unitary divine essence that was in constant movement. "Man" was at one with the cosmos and knowledge of astronomy, mathematics, music and metaphysics were part of the process of understanding the divine chain of being. The dualist notion of nature in the West is thus a result of a specific historical transition to modern modes of thought in which nature is separated from humanity.

Modern images of nature, generally dualist, can change human behavior radically according to specific spatial and historical circumstances: from exploiting forests for industrial production to protecting urban parks from development . However, all these images of nature have in common the idea that there is a being out there that escapes cultural, rational and ordered spaces as well as all instrumental calculation and technical efficiency. For these reasons, modern forms of governance imply a process of controlling, transforming, counting and dominating nature (from forest-gardens to biological instincts) in order to transform it into a subject of knowledge (a scientific category).

Nature as nonhuman or objective may also be possessed, which implies the notion of property or ownership. Consequently, modernity also allowed the natural to be abstracted from its local relationship with humanity to become a mobile commodity or a product (Ellen 1996a, 1996b, Ingold 1996a, 1996b, Blatter, Ingram and Doughman 2001). In the end, modernity assumed that nature could be "rationalized and controlled through laws, institutions, and organizational structures" (Blatter, Ingram and Levesque 2001:1). This is also a victory of humanism and individual human reason over nature insofar as nature can be "owned, moved, bought, and sold according to the dictates of individual interests and economic strength" (Blatter, Ingram and Doughman 2001:18).

This separation of culture from nature was reified within natural and social sciences, and it corresponded perfectly with evolutionary perspectives on scientific progress as a matter of demonstrating increasing technical ability to transform nature. Within modern tradition, nature became an object of knowledge with a telos that did not include or respond to human values, and that required that humanity demonstrate its power to negate nature's potentially threatening otherness and make it serve human values. In addition, the modern separation of humans from an eternally other nature implied that the natural is an inexhaustible object of control and domination requiring technical processes that must be produced, improved and reproduced in a progressive conception of time (Ellen 1996a, 1996b, Descola and Pálsson 1996).

Recent developments in ecology, genetics, biology, ethnobiology, social theory (poststructuralism and postmodernism), epistemology, sociology of science, social ecology, ethnography and biotechnology have helped to erase the borders between nature and culture and have given rise to the notion of nature as multiple and culturally constructed (Haraway 1991, Strathern 1992, Ellen 1996a, 1996b, Descola and Pálsson 1996, Escobar 1996, 1999, Blatter, Ingram and Doughman 2001, Blatter, Ingram and Levesque 2001). Among all these reconsiderations of the nature/culture separation, anthropologists, because of their relations with other cultural traditions, have been important in bringing into social sciences and modern thought new conceptions of nature in which modernity's distinctions between nature and culture are inapplicable.

The interrelation of different conceptions of nature also implies the recognition of nature as a cultural construction that changes according to specific historical situations and locations. People experience and conceptuaize, which is to say "construct," nature differently according to their particular social and cultural practices and institutions, material contexts and ideologies and morality.

Specifically situated practices and conceptions establish diverse ways to perceive, use and interact with nonhuman entities. Gender, class, race, age, status and ethnicity may significantly affect these notions of nature. From this perspective, nature no longer exists in objective isolation from humanity, but it is instead assumed always to be in dynamic relation with human values. Such an interaction between notions of nature implies a struggle over meaning that is political. It also implies that, for example, environmental "management" policies cannot be neutral or objective, and that the benefits and costs of environmental changes to humans are mediated through the unequal relations of power evident in our societies (Moore 1993, 1996, 1997, Escobar 1996, 1999, Peet and Watts 1996, Bryant and Bailey 1997).

Conceptualizations of nature as a cultural construct are now considered a taken-for-granted truth. However, this involved a long process of transformation from the hegemonic western notion of the nature/culture object/subject dichotomies to hybrid notions of quasi-objects and quasi-humans, and from nature as an apolitical entity to nature as a cultural construct with political implications (Haraway 1991, Strathern 1992, Latour 1993, Cronon 1995, Descola 1996, Ulloa 1996, Stone 1996, Milton 1996, Descola and Pálsson 1996, Little 1999, Brosius 1999a, Gragson and Blount 1999, Nazarea 1999, Ulloa 2001a, among others).

Contemporary environmental discourses often differ radically because some continue to use notions based on the traditional modern opposition between nature and culture and others use multiple discursive perspectives and argue for a new, nondichotomic relationship of nature and culture. Consequently, some environmental discourses are based on the conceptual distinction between nature and culture derived from the social and natural sciences, while others have been nurtured by new perspectives that promote new relationships with nature. Within this diversity of positions and tendencies, two main visions stand out: an eco-sustainable modernity and an alternative holism.

An Eco-sustainable Modernity

The proponents of this tendency propose global environmental solutions to the consequences of chemical fertilizers, transgenic foods or the slaughter of seals or whales; but they do not confront in a direct way the social, economic or political aspects that have generated these situations. This position does not question the current relationships among humans and their environment, which are based on the modern notion of nature and culture; neither do they question the current economic and political system. This tendency includes environmental movements that employ a so-called "shallow ecology" that is only concerned with the problems of contamination and shortage of resources, and not with the general causes of social, economic and environmental crises in the global context. Such "shallow ecology" assumes the modern notion of nature as an external and oppositional entity relative to humans. Following this notion implies that environmental solutions will focus on managing the biophysical environment and not on managing the interrelation of societies with their environments.

This anthropocentric perspective assumes that although the relationships of nature and culture are problematic, they can be resolved with more scientific management of ecosystems, circulation of and access to information, democratic participation and new market regulations and incentives (Anderson and Leal 1991, Williams and Mathenny 1995). Under these measures, it is possible to reach economic growth based on a wise use of natural resources (sustainable development) that conforms to the modern ideas of the domination and protection of nature. It also assumes that modern ideas of the management of nature will be the basis of global eco-governmentality.

This approach displaces environmental problems from broader political debate and places them under control of managerial programs. Thus, a new domain of knowledge called "ecosystem management" has appeared. This approach, for example, transforms the notion of a park as a specific space to preserve wilderness through the notion of an ecosystem that includes the interrelation of biological and social entities. However, this new notion of ecosystem management places the management of nature on a continuum with all the other ways human's control and discipline human and nonhuman bodies and practices, for example, agricultural practices and the industrial use of natural resources (Fairhead and Leach 1995). This new view of ecosystem management idea has attracted international allies who employ its language of extinction, depletion of species and natural resources and participation and community-based management in a discourse of development that retains many of modernity's assumptions regarding nature.

Brosius (1999) describes how, in 1997, in the Malaysian rain forest, the Penan people began political actions against industrial logging methods. The Penan's political actions found support among national and international NGOs that helped to locate the indigenous' claims within the international arenas. During more than 8 years, the Malaysian state, international NGOs and multilateral institutions struggled to resolve the logging problem as a problem of ecosystem management. In the end, the announcement that the Malaysian state would implement sustainable strategies for management of timber production "solved" the problem. Though the Penan people's ecological practices and conceptions of nature gained recognition around the world, the solution did not include their ecological practices or conceptions of nature. In fact, it all but guaranteed their destruction. According to Brosius, the Penan case is an example of the institutionalization of environmental problems that displace matters of political negotiation to the realm of management and technical solutions.

In a similar way, the agreements of Rio (1992) and the goals of the Kyoto Protocol (1997) related to environmental problems have not been pursued because they imply changes in the modern conceptions of economic production and consumption that dominate western thought; and environmental proposals such as decreasing pollution or swaps for nature (pollution credits) have been recommended without questioning the western role in producing the environmental problems that have given rise to such proposals for global management.

An Alternative Holism

In the second general trend within global environmental discourses, environmental problems are seen as a result of the "governmentality of the current economic and social regime [that] enforces its destructive disposition on things and people in the environment" (Luke 1997:196). Therefore, activists of this kind of environmentalism propose a new governmentality that includes nature as an equal. Furthermore, this vision promotes the life of future generations and the right to life of nonhumans based on communal and nonanthropocentric values (Dobson 1989, Williams and Mathenny 1995, Peet and Watts 1996, Luke 1997, 1999, Escobar 1998). Consequently, this trend supports a new environmental regime in which local communities develop their own social, ecological and cultural interests according to their specific environmental and political situations, and these local communities have to base their environmentalism on mutual responsibility and sharing. Thus, every community member has to share in the actions and collective decisions that construct the institutions of communal democracy. This communal democracy seeks to reinforce "a culture of fairness, toleration, and humility as they develop together with other communities seeking similar ends," while acting in reciprocity with nature (Luke 1997:205). In a similar way, Blatter, Ingram and Doughman (2001) argue that the interrelation of premodern, modern and postmodern ontology implies new ways of thinking about transboundary natural resources and, by extension, new conceptions of human/nonhuman relationships.

In this context, we can locate diverse philosophical positions: anti-industrial romanticism, anti-modernism, spiritualism, social ecology, populism, ecofeminism, among others, in addition to indigenous peoples' worldviews. Indigenous peoples around the world have taken a leading role in the holistic movement by promoting their conceptions, perceptions and practices regarding nature as alternative responses to current environmental models. From the perspective of the nonindigenous actors in alternative holism, nonwestern systems of knowledge, specifically indigenous knowledges, offer valuable alternative approaches to the environmental crisis because their visions of natural and social relations are reciprocal rather than hierarchical. For example, in some indigenous cultures, nonhuman beings have human behaviors, and they are regulated by social rules. Thus, the relations among humans and nonhumans are in constant transformation and reciprocity. These views arise from complex conceptions of nature that do not correspond to the western categories—although they have had longstanding interactions and interdependencies with those categories. Their categories and conceptions articulate nature as an integral and equal element of human identity and its social and cultural traditions. In other words, the indigenous worldview is that of a dynamic process that does not reproduce the rigid dualism and hierarchy of nature and culture. This perspective fits well within the biocentric paradigm. However, such a western view of indigenous peoples also invites stereotypes of them as "noble savages" located in ecological paradises separate from the environmentally fallen West (as we will see in following chapters).

Some groups in the environmental movements, drawing upon indigenous and grassroots practices, propose changing western society's relation to the environment and forming new patterns of consumption and production. Accordingly, some international NGO's and grassroots organizations are producing counterdiscourses that propose a new relation between humans and nonhumans that reshape and rethink predominant concepts of nation, citizenship, development, democracy and nature.

These proposals are rooted in notions of social and environmental justice. The movements denominated "radical environmentalists" and "pro-environmental justice" center their demands on changing and improving human ecological practices and conditions (waste disposal, pollution, toxic materials, among others) and questioning the social asymmetries and injustices that contribute to environmental damage. Movements like these oppose big development projects and they defend the rights of local residents. Although they assign great value to an equitable relationship with nature, they accept new technologies insofar as they serve as means to reform social relationships and create democratic spaces. The more radical proposals within this tendency come from the well-known movements such as "Deep Ecology" and "Earth First!" The group Deep Ecology arose in 1972 with a proposal for wilderness protection, focusing its actions on what it denominated a biospheric egalitarianism in which the humans have the same rights as other species. Following this tradition, Earth First!'s members promote an ethics that prioritizes the interests of nature over the interests of humans. However, these movements have been criticized because they do not consider social problems and planning major parts of the environmental crisis and because they focus primarily on the idea of wilderness.

THE GLOBALIZATION OF ENVIRONMENTALISM (ECO-GOVERNMENTALITY)

Although environmentalism has produced various and often contradictory answers to the problem of the global environmental crisis, they remain unified in their belief that an answer is needed. The quest for this answer has created movements, policies, and government agencies whose activities purposefully or accidentally transcend local interests and conceptions and generate in this way a series of interdependencies between local and global contexts.

The globalization of environmental concern began in the 1960s. At that time, resolving human problems of pollution and extinction required new national solutions, new expert knowledge and new environmental practices. It also required international participation. Old and new problems became problems to human life on a larger scale than ever before (global warming, depletion of natural resources, population growth), and this helped to place environmental issues on the global agenda. This increased attention in large part resulted from the impact of The Limits of the Growth Report (1972), The Global 2000 Report (1980) and The Brundtland Report (1987). Consequently, depletion and extinction of natural resources, pollution and population growth were identified as the key physical symptoms of the emergent global environmental crisis. These reports highlighted global and local threats to the environment as threats to the continuation of life on earth. Although such global-environmental discourses presented environmental degradation and pollution as a result of human activities, they often did so without giving thorough explanations of what causes those activities to be so harmful or how to mitigate their effects. Nonetheless, they did formulate the degradation of the global-environment as a problem whose solution belongs to all the citizens of the planet.

However, more recent theories of political ecology have begun to focus on unequal social power relationships as causes of the environmental crisis, thereby suggesting potential social and political solutions to the global environmental crisis. Many experts have begun to conceptualize a new object of knowledge called "the environment" in a manner that does not assume the management of nature in terms of parks and natural reserves designated for human enjoyment or as a public resource that the state manages to promote economic growth. This new idea addresses the reproduction of life on a planetary scale that shifts concern from national interests in conservation, recreation and development toward the necessity of biological diversity to the maintenance of life, human or nonhuman. Although this approach to the environment focuses its concern on human life, it also emphasizes concern for nonhuman species. Human life thus has a new meaning because it is not only responsible for the reproduction of society itself, but also for the reproduction and diversity of nonhuman species. Thus, the environment has become a new object of knowledge that transcends strictly human interests and calls for a special global perspective and technical governance. In this sense, this is the beginning of "geopower" [I am using Luke's (1999) term].

Such an environmentalism can be considered the birth of a new discursive formation according to Foucault's concept of discourse, and so it has produced a group of statements that provide a language for talking about (a way of representing the knowledge of) "the environment," the term that unifies these languages and statements as elements of a particular historical and conceptual moment. Consequently, environmentalism in this sense encompasses a variety of practices and conceptions (recycling, green consumerism, frugal ways of living), expert knowledges (social ecology, forestry, bioprospectors, ecosystem managers), texts (information about global warming, biodiversity, rain forests, endangered species), technologies (GIS, sonography), policies (sustainable development, international treaties related to biodiversity, global surveillance), objects (green products, eco-art), representations (recycling signs, organic labels), and subjects (ecological people, pre-industrial people). Insofar as these various disciplines and activities share the same discursive assumption (environmental crisis) they belong to what Foucault would characterize as the same discursive formation. In this way, the environment now appears as the "truth" that regulates and synchronizes the reproduction and continuation of human and nonhuman life.

Global environmental awareness has helped to transform citizens into green-citizens through normalizing processes. Thus, around the world, we implement disciplines through daily practices such as recycling, green consumerism (recycling products and biodegradable-commodities such as soaps, natural bottles of water, organic fruits), school programs (earth day, clean your beach, mapping endangered species) and social processes that involve ecologically sound standards. At the same time, schools, neighborhoods, marketing strategists, economic think tanks and satellites systems have become parts of the surveillance processes that guarantee the production of green-bodies: in other words they produce the information and ecological awareness that regulate and synchronize the daily practices of individuals.

At the same time, the environment as a "specific way of talking" has produced other specific subjects (*ecological natives*) that require special representations as pre-industrial people or indigenous people to transform them into personifications of or participants in such environmentalism. In this way, an "ecological other" has been incorporated into the global environmental discourse in a way that conflates indigenous peoples and environmentalism. Although this "other" may often be little more that a western stereotype, it does have advantages insofar as it allows indigenous and western peoples to formulate new ways of talking about "nature" and about relationships among humans/nonhumans that promote different ways of using and managing the environment.

Furthermore, it appears that the relationship of indigenous peoples and ecology allows new practices and conceptions about democracy, citizenship and participation that challenge modern notions. Moreover, the new global environmental discourse, particularly the idea of sustainable development, promotes protests by indigenous peoples and local claims for self-determination. In fact, local protests against transnational corporations responsible for environmental destruction, have resulted from this environmental discourse. Thus, in the symbolic terms of the environmental or ecological imaginary, indigenous peoples have already won the right to implement in their territories their own legal and managerial systems and to employ their natural resources according to their own cultural practices and local knowledges.

More practically, new international legal instruments related to indigenous peoples' rights (ILO-169, CBD) demonstrate a new vision of indigenous peoples in relation to their territories and natural resources. The ILO-169 states that one of its purposes is "to contribute to stimulating their economy and consolidating their control over their territories and/or their natural resources: to improve and enable their productive capacity and their transformation in the context of healthy ecology" (Annex A: 1). Environmental and indigenous peoples' policies thus arose together and combined in a way that links local and global concerns through programs and actors working together in a specific place: indigenous peoples' territories.

However, disciplines that construct "green bodies" most often do so in a manner that reproduces western values through international technologies of global environmental security. In fact, new supranational environmental institutions such as the CBD and the GEF have begun to regulate the environment in economic terms and under a capitalist framework of international markets in a way that constructs biodiversity as "a world currency" (McAfee 1999). These generally neoliberal regulations are designed to defend the planet and protect natural resources for the good of "the New World Order."

International surveillance through the World Watch Institute, the World Wild Life Fund (WWF) and the World Conservation Union (IUCN) constitutes an international gaze that covers the entire planet (an eco-panopticon[3] according to Luke 1999). In this way, soil erosion, extinction of species and global warming fall under the gaze of scientific knowledge. These calculations of species, resources and degradation provide the basic data for implementing international policies and research priorities. International NGOs and governmental programs have also become part of these processes of surveillance, and thus from a global perspective they detect local environmental dangers and threats to humanity. This information also contributes to the empowerment of the western environmental gaze and, often inadvertently, to that of neoliberal economic programs. So, too, the environmental networks (NGOs, grassroots and academic communities, among others) have contributed to the globalization of western environmental interests to the extent that they tailor their projects to the economic interests of the wealthy western sponsors that provide the money to realize them.

Such a western-dominated eco-governmentality determines daily environmental practices in a manner that places indigenous peoples in a framework of new processes of production and consumption that at once employ and transform indigenous ecological practices and knowledges. However, according to McAfee (1999) and Gupta (1998), because the recognition of indigenous contributions is linked to their economic potential, indigenous peoples' knowledges and territories are not being recognized for their inherent cultural and environmental value, but rather for their market value in the West. So it is that when nature becomes a global currency (Gupta 1998, McAfee 1999, Sachs 1999), the practices and knowledges of indigenous peoples gain recognition only because they are valuable in this new "free" eco-market (see chapter 8).

Consequently, indigenous peoples have to implement their practices under national standards of ecological safety that reproduce international patterns of sustainable development based on notions of the "free" market (McAfee 1999). As we have seen in the case of the Penan people of Malaysia, sustainable development projects have been introduced in "third-world" countries that impose a global management of natural resources without consideration of indigenous environmental practices and strategies. In some cases, the coalition of indigenous and environmental movements has had negative implications for indigenous peoples' autonomy within their territories and control over their resources. This is evident among those environmental movements with a biocentric perspective that pursue their goal of preserving wilderness without consideration of indigenous peoples' practices or strategies (Arvelo 1995, Varese 1995).

In the end, an environmental awareness that seems supportive of indigenous peoples can negatively affect and transform indigenous local and national practices according to the variety of nonindigenous concerns and interests that currently dominate the new transnational environmental discourse. In fact, this global approach causes changes in local practices, epistemologies and identities very similar to the changes caused by the western idea of progressive economic development or, for that matter, colonialism. McAfee (1999) calls these transnational processes of controlling biodiversity "green developmentalism."

In this sense, indigenous peoples are now called to give their knowledges (that were before ignored) to humankind as an expression of solidarity with "our" environmental goals. Thus, indigenous peoples have the historical responsibility of maintaining and reproducing the life of human and nonhuman species. As a result, their territories and the genetic resources within them become the reserve of humankind and indigenous peoples become the donors of this life to humankind. As "keepers of nature," they have the historical responsibility of protecting their territories and maintaining biodiversity without changing their cultural practices despite these new demands. The imposition of this new eco-governmentality and its eco-disciplines and discourses upon indigenous bodies and territories is largely meant to maintain standards of living in societies that do not want to change capitalist patterns of production and consumption. It is not clear how the indigenous will share in the benefits of their biodiverse "treasures." However, it is clear that the globalization of the environment described above has caused changes (too often negative) in the daily practices of indigenous peoples at the local level. Nonetheless, the cultural politics of indigenous peoples' movements demonstrates an awareness of this situation and a willingness to challenge and change the terms of this relationship.

Environmental discourses reflect, in general, the priorities and necessities of the environmental movements in developed nations. They are an answer to processes that differ from those generated in countries of the so-called "third-world." In fact, environmentalism is often considered a phenomenon of industrialized societies whose excess of consumer products has given rise to a postmaterial value system. For these reasons, some scholars argue that this environmentalism ignores the proposals of countries undergoing development. In these countries there is a different environmentalism which has been called, among others things, "the environmentalism of the poor." This approach combines environmental problems with issues of social justice or "the environmentalism of the third-world," which may also combine with ideologies such as eco-feminism and development. The activists of these tendencies want to rethink the idea of development by criticizing industrial and urban life and arguing for the rights of poor communities. In short, the emergence of environmental movements has to be analyzed according to the specific historical contexts in which they emerge and according to the necessities and interests that they address. Consequently, the interactions of indigenous movements and environmental movements in Colombia must also be approached as a multivalent and often contradictory process.

ENVIRONMENTAL AWARENESS IN COLOMBIA

In Colombia, the emergence of environmental movements as well as the notions of nature, environment and ecology, have been experienced and conceptualized differently according to particular social locations and political interests. Scholars, indigenous peoples, peasants, and more recently environmentalists have advocated various notions and practical relations to nature, but their common result is a growing environmental awareness within the Colombian national context.

Some scholars claim that environmental awareness in Colombia began during the 19th century with the naturalist tradition exemplified by two scientific expeditions: La Expedición Botánica and La Comisión Corográfica. For Márquez (1997), it was only in 1948 when Professor César Pérez taught the fist class in ecology (Universidad Nacional, Medellín) that environmental awareness began. This initial moment was followed by the activities of a priest, Enrique Pérez Arbeláez, who formed the National Institution of Sciences and the Department of Biology in the Universidad Nacional de Colombia, Bogotá.

During the 1960s, different scholars from the natural sciences made collections of plants and animals and pursued geographic investigations. This initial period promoted scientific research in which "expert knowledge" was basic to understanding nature. Developments in ecology and the expansion of scientific institutions (Sociedad Colombiana de Ecología in 1972, Ideade in 1985 and Instituto de Estudios Ambientales-IDEA in 1989, among others.) have continued since then. However, Colombia's environmental discourses have also changed in response to the influence of international ENGOs and social movements (indigenous peoples, peasants and Afro-Colombians) (Márquez 1997). During the 1970s, international discussions around the world regarding the environment (limits to growth in the Club of Rome in 1972, the Meadows report in 1970, and the environmental movement in the United States, among others) affected the Colombian context in two different ways: the introduction of international policies and the concept of ENGOs.

In this sense, part of Colombian environmental awareness is an expression of the international governmentality that situates Colombia as part of global political and economic regimes. In fact, in the 1970s, national policies began to incorporate the emergent international view of natural resources that had transformed them into environmental elements within a global conservation strategy. Thus, the national government formed national parks in order to conserve and preserve unique ecosystems. National policies were implemented to regulate the use of natural resources such as water, fauna and flora (Código Nacional de Recursos Naturales 1974), and created new institutions to implement these policies (Inderena, Sistema de Parques Nacionales).

These institutions implemented policies that promoted the conservation of natural areas, the restriction of use in some territories, the control and regulation of natural resources, the reintroduction of endemic species in "natural territories" and the creation of zoos and places for reproduction of wild species (botanical gardens, nurseries for wild animals). Similarly, international development programs and sustainable development policies have helped to reshape national notions of "nature." In fact, "third-world countries" had to include, according to international requirements, "development programs" for their environments in order to become "first-world" countries (or "civilized" in colonial terms). However, parallel to these national institutional processes, national nongovernmental organizations began to develop conservationist programs (Fundación Herencia Verde, and Cosmos, among others). Local ecological groups (Consejo Ecológico de la Región

Centro Occidental-CERCO, and Grupos Ecológicos de Risaralda-GER, among others) also began to link conservation to social activism. These NGOs and local groups offered a new perspective that combined conservation with social problems and protests against development. This social perspective promoted coalitions of these environmental organizations and indigenous, Afro-Colombian and peasants' movements in their efforts to link local social, economic and political situations to national and international strategies for the management of natural resources (Alvarez 1997).

These particular strategies contested institutional development and environmental programs such as resource extraction and the creation of national parks when they did not arise from local initiative or were imposed without local consultation or consensus. (Today 20 natural parks overlap traditional indigenous territories.) In addition, these strategies usually employed conceptions of development and management that did not conform to local practices. For these reasons, various Colombian social movements have challenged and transformed international and national environmental policies and caused them to consider, if not include, indigenous perspectives about "nature" in their plans.

In Colombia, indigenous peoples' conceptions related to territory and human and nonhuman relationships constitute one of the original modes of what we now call environmental practices. Moreover, the indigenous peoples' have given Colombian NGOs and local groups conceptual tools to fight for new relations between national society and the environment. For example, the indigenous peoples' conceptions about human/nonhuman relationships encourage a reciprocal relation that promotes a different understanding of development that seeks to maintain indigenous peoples' cultural practices by protecting the environment. For this reason, indigenous peoples' conceptions began to be identified with environmental discourses. The spread of indigenous environmental knowledges also occurred as a result of anthropological studies. Since the late 1960s, some scholars began to write ethnographic studies about indigenous peoples' ecological views and management of their environments (Reichel-Dolmatoff 1967, 1968, 1976, 1978, 1988, 1996, Hildebrand 1983, Reichel 1989, Correa 1990, Van der Hammen 1992, Mora 1995, Ulloa 1996, 2001a, 2002a, among others). The combination of global environmental concerns, the emergence of Colombian environmental organizations, ethnographic studies and indigenous peoples' social movements resulted in an increased awareness and recognition of indigenous ecological knowledges. This recognition subsequently found practical expression in the political discourses of governmental, nongovernmental and indigenous organizations (see chapters 5,

6 and 7). Other practical consequences include the increased number of NGOs, programs, institutions, policies and studies that addressed the relation of indigenous peoples to the environment that occurred in the 1980s. New ENGOs (Mayda, Penca de Sábila, Bacatá, Fundación Natura, Fescol, Fundación Pro Sierra Nevada de Santa Marta) led such environmental activism, and there was a corresponding increase in academic environmental programs in universities around the country.

All these actors (scholars, researchers, peasants, indigenous and Afro-Colombian movements, urban grassroots organizations, some public employees, universities and transnational NGOs) promoted different activities, interests and conceptions with respect to nature. However, some scholars (Palacio, Alvarez, Carrizosa) argue that the interrelation of all these positions resulted in the emergence of a national environmental movement (Palacio1997). Some of these actors contributed to Colombian environmentalism through participation in three important national meetings: Ecogente-Pereira (1983), Cachipay (1985) and Guaduas (1992).

These diverse actors and agencies also began to articulate themselves under a national organization, Ecofondo (1991), which is an entity that forms part of both state and civil society. In this way, there is some degree of national coordination of the environmental movements of civil society; however some movements are dispersed and do not have a centralized organization. The main goal of these actors and agencies is the protection of natural resources; however they use different strategies to do so (from protection of wild animals to promotion of sustainable development). They also combine these strategies with particular interests such as human rights, peace, ethnic rights, and the search for alternatives to western-oriented development (Carrizosa 1997).

Palacio (1997) argues that the environmental movement in Colombia has four general aims: conservation; linking social movements to environmental relationships; participating in international initiatives; and seeking financial, technocratic and intellectual support. Biologists and naturalists who seek to preserve wilderness take the conservationist approach. In the second approach, some NGOs and governmental programs address the proposals of social movements (indigenous peoples, Afro-Colombian and peasants) in an effort to rethink human and nonhuman environmental relationships. In contrast, the approach that follows international initiatives reproduces international concerns through dependence on international financial support. People who take this approach promote, for example, debt-for-nature swaps. Finally, the technocratic approach attempts to strengthen ENGOs at political-institutional levels in their practical efforts to ameliorate environmental problems.

These different visions began to consolidate after the declaration that Colombia was a country of high biodiversity in Rio-1992. That declaration implied implementation of global sustainable development programs. It also was a call for unifying ecological positions and constructing a sort of national identity based on this ecological wealth. In fact, most of the territory of Colombia was classified as a *biodiversity hot spot,* particularly the Amazonian basin as a *tropical wilderness area* with exceptionally rich animal and plant species, and the Chocó region on the Pacific coast with its 2,250 unique plant species.

Colombia has thus been incorporated into international governmental processes that involve a specific eco-governmentality and follow a specific "truth" of what has to be done in relation to the environment. In fact, in 1993, the national environmental institution (Ministerio del Medio Ambiente) was formed with different centers for pursuing biological research on biodiversity (Instituto Von Humbolt, Instituto Von Neuman) that require the production of specific forms of scientific knowledge. In addition, the GEF gave financial support to pursue research on national biodiversity. In response, Colombian environmentalists began to calculate, classify and organize a scientific environment consistent with the international standards (largely defined in the West). However, this does not mean that other visions of nature and environmental solutions cannot be included. In fact, paradoxically, it is in this new international and national context of environmental awareness that alternative ecological proposals find recognition and representation. In short, it is largely because of international influence on the Colombian government that indigenous knowledges, territories and resources have gained importance in Colombia.

Indigenous peoples must often modify their plans and activities in accordance with national standards of ecological management that mirror international schemes of sustainable development based on notions of the western "free market" (McAfee 1999). This interrelation of distinct notions of nature implies a political process of struggle over meaning. It also implies that environmental policies are not neutral and that unequal relations of power mediate access, benefits and costs with respect to the use of indigenous "resources." Moreover, because the indigenous peoples' territories have a great biodiversity, as well as minerals and oil crucial to the international economy, national and transnational business interventions often take place against indigenous wishes in the name of "rational" use and development (for example, OXY Oil Company exploration in the Uwa territory, or the construction of a hydroelectric in the Embera-katio territory). These territories are also strategic for paramilitary and guerrilla activities as well as illicit crops as long as they remain outside of state political control, which has increased violence against indigenous peoples and decreased their control of their territories. As a consequence, Colombia's indigenous peoples suffer a constant threat of displacement from their territories because of political violence, drug traffickers and transnational and national economic interventions (Quiñones 1998, Roldán 1998, ONIC 1998).

Nonetheless, indigenous movements' cultural and environmental politics are constantly challenging and reshaping these situations. In fact, they have presented different proposals to help in the national peace and development processes. They base their proposals on the constitutional recognition of their right of self-determination, their cultures, their knowledge and practical relations with nature, and their territorial claims (and they include Afro-Colombians and peasants in their plans) (ONIC 1998, Achito 1998, Alonso 1998).

In the last five years, the activities of environmental movements in Colombia have focused on several topics: the policies related to the CBD, the Kyoto Protocol, globalization, relationships with institutional political campaigns, the management of resources, the formation of a national environmental movement, activism linked to the demands of Uwa and Embera-katio peoples, and discussions on transgenic foods and access to genetic resources, among others. All this demonstrates the multiplicity of indigenous and nonindigenous participants and environmental issues in the dialogues that result in environmental decision-making. Nonetheless, this diversity shares the comon issue of the environment. Therefore, it is difficult to determine if there is or is not a unified national environmental movement in Colombia.

In the beginning, the objectives of Colombian environmental movements were to create national environmental awareness, influence politicians and situate the environmental crisis in the national and international scenarios. Those goals were accomplished and appear practically in the current development of environmental activities in the academic environment and public sector (forums, seminars, congresses, work groups, among others). Moreover, their influence in political arenas could be seen as the triumphs of both the Colombian indigenous and environmental movements. However, politicians have incorporated indigenous and nonindigenous environmental interests under the western logic of sustainable development, which demonstrates the conceptual difficulty of fashioning an equitable and noncoercive environmentalism in which the values and plans of indigenous peoples, alternative environmentalists and dominant western interests coincide.

INDIGENOUS PEOPLES OF THE SNSM AND ENVIRONMENTAL POLICIES[4]

The history of Colombian state policies on the SNSM's environmental problematic cannot be separated from modern thinking about environmental management and international policies, particularly the convention related to plants, animals and landscapes ratified in Washington on January 17, 1941. Through this treaty, American countries established agreements concerning the creation and definition of protected areas in order to preserve spaces of aesthetic, historical and scientific importance. In the laws and decrees of following years, (Law 52 — 1948, Decree 438 — 1949, Decree 2963 — 1955) the consolidation of forest reservations began, starting with the creation and delimitation of the Macarena reservation.

International forestry policies were also formulated in Law No. 2 dated February 27, 1959, which clearly sets out policies on the nation's forestry economy and the conservation of renewable resources. It established several zones as protected forest areas and general reservations within rainforests (Pacific, Magdalena River, the Sierra Nevada de Santa Marta, the Sierra Nevada del Cocuy and the Amazonia). These were conceived in terms of forestry science as zones needed to protect soil, water and wildlife and for potential commercial development under management plans submitted to the Ministry of Agriculture. In parallel, Article 14 of the same law also established national parks.

In the 1960s, the national environmental policies started to respond to international interest in conserving natural resources through a global strategy. The Colombian government established more natural parks specifically to conserve and preserve globally unique ecosystems. Later, several national policies were implemented to regulate the use of natural resources, such as water, plants and animals (the National Code of the Natural Resources 1974), and new institutions were created to implement these policies, such as The National Institution for Natural Resources-Inderena, National Park System, among others.

At the beginning of the 1990s, the Colombian national government implemented sustainable development policies in response to the Brundtland Report (1987) and associated global policies addressing sustainable development. In the context of the state's view of environmentalism, the report strengthened the argument for the "rational" development and use of resources and gave a globally recognized "environmental" rationale for national proposals regarding the imposition of biodiversity and sustainable management practices upon indigenous peoples and territories.

In the 1990s, this interrelation of global and national policies affected and changed the SNSM's environmental management. Since then, there has been a political agreement among all social actors who live there regarding the need to develop environmental programs (see table 3.1). However, the implementation of global environmental policies through national environmental law has been given priority over any regional or local environmental agreements or policies. Environmental policies for the SNSM thus conform to national environmental principles founded upon global guidelines for sustainable development.

Colombian Environmental Policies

Highlighted in the following sections are the most outstanding aspects of Colombian environmental policies, and regardless of the fact that they are presented separately, they are interrelated parts of a coordinated policy.

National Environmental Policies Framed within Global Contexts

The Rio declaration in 1992 constructs a transnational agreement related to environmental policies which addresses a "global crisis affecting national economics." This declaration acknowledges specific *threats:* climatic changes, loss of biological diversity, soil deterioration, deforestation, rainforest degradation, continental and sea-water pollution, destruction of the ozone layer, and the accumulation of persistent organic forms of pollution (MMA 2000). The Collective Environmental Project (*Proyecto Colectivo Ambiental*-PCA, MMA 2000) reflected this global trend when it announced that:

> Obligations are derived from a commitment to the planet and our country's environmental health, as from the common but differentiated principle of responsibility. Rights derived from environmental services that the country renders to the planet, through the ecosystems located in our territory that, due to its privileged wealth, is a concern for humanity. (PCA, MMA 2000:19)

Because of such global environmental concerns, multinational agreements have been formulated. Colombia has signed approximately 105 international agreements, among which we can point to the Rio-Declaration-92, the CBD (Law 165 - 1994), and the Decision 391 in Cartagena (1996) on access to genetic resources. These environmental policies between nations are designed to consolidate alliances and coordinate actions that address global "threats" (MMA 2000).

Sustainable Development as an Objective[5]

Through national policies framed in a global context, "sustainable development" became part of all governmental and nongovernmental environmental discourses, and the use of this term gained much of that importance as a result of Rio-92. For example, Colombian's Law 99 of 1993 states that the national "policies and regulations of recovery, conservation, protection, ordering, management, use and benefit of natural resources and the environment will serve to guarantee sustainable development." In addition to sustainable development, the concept of sustainable use is incorporated as an environmental conservation policy (Law 165-1994). Therefore, national environmental policies have to be implemented within the terms of sustainable regional development plans. Article No.31 of the VI Chapter of Law 99, referring to Regional Autonomous Corporations (CARS), establishes the relationship of sustainable development and "*ethnic groups.*" CARS put in motion the coordination of authorities from these regional corporate groups with programs and projects for sustainable development and environmental management. They have been given the difficult task of managing both the commercial use of natural resources and the conservation of the environment.

The Environment as an Economic Value

In this context, the environment is a national and global economic resource—a concept that derives from the concept of sustainable development. In the national context, governmental and nongovernmental institutions began to consider the environment a public good whose deterioration affects national economies. Biodiversity and environmental services began to be thought of as "economic values," and as a consequence these policies began to assess the value of the "national environment" in relation to global economy and international environmental and economic policies. This new perspective has generated different views of the environment that have had practical results in the form of environmental services, green markets, economic incentives and compensations, among others. This economic perspective also appears in strategies formulated in environmental policies designed to generate production for local and foreign markets. In the PCA (MMA 2000) one of the priority programs is to contribute to environmental sustainability (objective 3) through green markets, with the objective of "encouraging environmental services and products and increasing competitive ecological services offered in local and foreign markets."

Work on Strategic Ecosystems

In Colombia, specific areas began to be thought of as ecoregions, defined according to biological and political criteria and their national, regional and local importance. An ecoregion is defined "by direct or indirect value, immediate or potential, represented in services provided to man." The assumption is that this approach to ecoregions guarantees that the regional realization of guidelines and directives of the collective environmental project effectively contributes to strengthening the National Environmental System-SINA. The reference to "value" also refers to economic development as a goal of that system.

The coordination of these ecoregions has to respond to two major considerations: water conservation and commercial production. On the one hand, conservation is related to actions aimed at improving water quality and guaranteeing its regulation. Production, on the other, is the coordination of actions designed to identify goods and environmental services and to generate productive alternatives and open green markets for products derived from the regions' biodiversity (MMA 2000). The SNSM is such a strategic ecoregion because, in addition to meeting biological and political requirements to be classified as such, it has an evident problem of water shortages that requires management within this environmental strategy.

Territorial Ordering As an Environmental Planning and Management Instrument

Territorial ordering refers to the state's power to direct the use of territory in order to promote environmental sustainability at the local, regional and national levels. The CPC-1991 recognized indigenous peoples' lands as territorial entities (ETIS) over which they have autonomy to manage their activities according to cultural practices and interests (articles 286 and 287). Regardless of such constitutional recognition, the ETIS have not been regulated. It is the responsibility of all different territorial entities' authorities (departments, districts, municipalities, indigenous peoples' territories and those of Afro-Colombian peoples) to prepare territorial ordering plans (Plan de Ordenamiento Territorial-POT). In this haphazard way, indigenous peoples have been introduced into the modern concept of territorial planning (see Comaroff and Comaroff 1997).

Inter-institutional Coordination with All Social Actors

This means coordination and management of territorial entities under environmental jurisdiction and by means of different social and institutional agents in processes that organize and provide coherence to management strategies in environmentally defined territories.

Social Perspective within Environmental Policies

Colombian environmental policy, as previously mentioned, attempts to alert different social agents that environmental problems have to be addressed in regional and local settings and how the environmental problems affect national society as a whole, especially its various economic sectors (MMA 2000). Current environmental management policies also emphasize conservation strategies as an important means to resolve social conflicts. They aim to generate economic options for people, to increase the quality of life and alimentary security, and to guide development processes. The basis of this policy derives from the CPC-91 which emphasizes the idea of a *state of law* (for all the nation) that promotes decentralization, participation, fundamental and collective rights of security for citizens, establishment of supervision mechanisms for the state's management programs and social responsibility for the environment's protection.

Additionally, the national environmental policy is based on the recognition of ethnic and cultural diversity formalized in CPC-91 and the ratification of ILO-169. Specifically regarding natural resources, CPC-91 establishes that their exploitation will be accomplished without any deterioration of cultural, social or economic integrity of indigenous peoples (article 330). It also establishes that local governments will take any measure needed in cooperation with indigenous peoples to protect and conserve the environment. Likewise it establishes that the government has to consult indigenous peoples in the event of any prospecting or commercial exploitation of the natural resources existing in their territories. Decision 391 of Cartagena (1996) also states the need to "recognize the historical contribution of indigenous, Afro-Americans and local communities to biological diversity, its conservation and development."

National environmental policy is also linked to the recognition of indigenous peoples' environmental knowledge and practices. The PCA reflects this environmental perspective "in its formulation and execution as a participating process grounded on the environmental, cultural and social specificity of the different regions" (2000:13). The PCA supports the restoration and protection of nonwestern knowledge systems, traditional practices and innovations in order to assure their contribution to the formation of regional development models.

In turn, the new policies directing the national system of protected areas[6] promote the conservation of both natural and cultural resources such as those maintained by indigenous peoples and ethnic groups in general. These policies specifically recognize the contribution of different indigenous and Afro-Colombian communities in environmental management and how they may contribute to generating new local and regional development alternatives. The document called *Parks with People* (UAESPNN, 2002) explicitly addresses recognition of indigenous knowledges and aims to guarantee the "social and cultural sustainability of protected areas through participation processes, defining and integrating management processes and managing these areas in a transition strategy that will consolidate the protected areas and allow it to solve the problems concerning area's of unlawful use and occupation, and thus resolve conflicting factors" (MMA 2000:69). Based on this perspective, one of its objectives is the consolidation of social and institutional processes that affect conservation.

The relationship of indigenous people to biological diversity and sustainable use thus arose in the context of these environmental actions and policies, and this is how cultural issues became an element in environmental management, sometimes as a resource needing protection and sometimes as an agent whose autonomy and participation directly influences the form and impact of such management. This objective appears in the Collective Environmental Project (PCA) from the Ministry of the Environment (2000), which states that national environmental policies should include local participants in "collective work related to *sustainable development* [which] has to be based on *social and environmental potential* framed in terms of *participation* that leads to *ecological respect* and *preserves* the environment, as a *national and global resource* for present and future generations" (MMA 2000:11). In other words, it aims to integrate local human and nonhuman relationships in a development plan that links the local to the global *(that is to an eco-governmentality in its hegemonic dimension).*

Indigenous Peoples Environmental Politics

In the SNSM, indigenous peoples' ecological practices are a result of spiritual activities that incorporate their knowledge of ecological interactions, the natural history of species, abiotic components (climate, soil), and astrology, among others. The spiritual perspective of indigenous peoples promotes the use of resources according to ecological conditions and the need to distribute the impact of human activities across resources, habitats and geographical areas. The ecosystemic view of indigenous peoples is evident in their interaction with "nature" in the territory delimited by the Black Line that extends from the lowlands to the highlands. For example, they conceive of the water basins as a whole, and they base their activities on the interrelation of the different climes of the SNSM. For them, this interaction between the human and nonhuman constitutes an integral spiritual activity (see table 7.5 and chapters 3 and 7).

However, indigenous peoples are now incorporated within national and global policies that have already established priorities related to environmental concerns in their territories: water conservation, conservation of biodiversity and the formulation of economic strategies for managing natural resources. Because those policies impose western processes of participation and decentralization to generate answers to the question of indigenous peoples' rights (*neoliberal multiculturalisms*), indigenous peoples have to adapt their ecological practices to the nonindigenous context of national and transnational environmental policies. Indigenous peoples, therefore, have to begin to consider how to formulate a *coherent and modern strategy* of managing natural resources under the modern categories of nature and culture and progressive economic development.

CONTROVERSY BETWEEN INDIGENOUS PEOPLES' POLITICS AND GOVERNMENTAL POLICIES

Various factors affect, limit and, in general, jeopardize discussions concerning indigenous peoples and environmental management of the SNSM (as we have seen in previous chapters). The main one is the fact that the indigenous integrate the environment to their cultural, social, economic and political interests. For example, among the indigenous peoples, environmental management is related to political autonomy and self-determination rather than purely biological or economic concerns, as it often is for the nonindigenous.

In this context a variety of agents (see table 3.1) have different, and sometimes opposed perceptions of the SNSM's environmental management and the benefits that they expect to obtain from its natural resources. According to national policies, participation, consultation, equality and autonomous processes have to take into account within the SNSM's environmental management plan the interests of indigenous peoples and governmental and nongovernmental institutions, among others. The management of water resources is a common element that concerns all local agents and that likewise coincides with a priority of national and global environmental politic, but for often clearly different reasons (see tables[7] 4.1 and 4.2).

Table 4.1. Environmental Problems in the SNSM's According to Different Actors and their Priorities

GOVERNMENTAL INSTITUTIONS	ECONOMIC SECTORS	INDIGENOUS PEOPLES	PEASANTS
Armed conflict and violence.	Armed conflict and violence.	Traditional territory loss in lowlands of the SNSM.	Armed conflict and violence.
Different governmental criteria to administer the SNSM.	Production cost overruns.	Loss and desecration of sacred sites.	Deforestation and lack of water.
Fragmented and contradictory management.	Water supply problems.	Pollution from deforestation in lowlands of the SNSM and construction of development projects.	Absence of state presence reflected in lack of education, employment and health services.
Lack of financial, technical and human resources.		Treasure hunting at archeological sites.	Land distribution.
Land distribution.		Institutional lack of coordination.	
		Armed conflict and violence.	

From the evidence of the water management issue, it could be concluded that governmental policies and indigenous peoples' politics have a common concern: to conserve the Sierra Nevada de Santa Marta. Despite these and other points of agreement in the joint environmental management proposals (see table 4.2), basic conceptual differences remain between governmental and nongovernmental institutions and indigenous peoples' perspectives.

Table 4.2. Common Aspects and Perspectives Related to the Environmental Management of the SNSM between Indigenous Peoples and Governmental Institutions

ENVIRONMENTAL ASPECT	GOVERNMENTAL PERSPECTIVE	INDIGENOUS PEOPLES' PERSPECTIVE
Need for a project to define SNSM's management and use	Conservation (biological and cultural diversity).	Strengthen conservation, territorial unity and cultural permanence.
SNSM's conservation leads to a world benefit	World strategic region due to its biodiversity. Biosphere reserve.	SNSM is the world's heart. On it depends the rest of the world's well-being.
SNSM as a territory	Strategic Ecoregion defined by natural and political elements.	Ancestral territory delimited by the Black Line that delimits interrelation between the spiritual and the material.
Obligation to conserve the SNSM	Granted through CPC-91 and Law 99-93.	Granted through the Original Law.
Necessity of social agreement	Institutions, other users (POTs, PDS) where all the sectors' interests are considered.	Institutions and other users of the SNSM (CARs, UESPNN, SINA and development of POTs) where indigenous peoples' proposals have to be reflected.
Recognition of in digenous peoples' traditional management as the most adequate	National Natural Parks' recognition of indigenous peoples' environmental authorities.	Ancestral knowledge and management according to the Original Law.
Recognition of different cultural realities	CPC-91, Law 21, Policies of Parks with People, Research Policies, Indigenous Peoples' Policies.	Stated in draft documents of SNSM Indigenous Peoples' Policies.

(*continued*)

Table 4.2. (*continued*)

ENVIRONMENTAL ASPECT	GOVERNMENTAL PERSPECTIVE	INDIGENOUS PEOPLES' PERSPECTIVE
Intercultural relationships based on autonomy, participation, consultation and common consent	CPC-91, Law 21, Law 99 of 1993, Policy of Parks with People, research Policy, participation and agreement policies with in digenous communities.	Stated in draft documents of SNSM Indigenous Peoples' Policies.
Recognition of autonomy and jurisdiction of indigenous peoples' territories	CPC-91, Law 21, Law 99 of 1993, Decree 2164 of 1995.	Through the Original Law.
Water as the core of territorial ordering	River basins.	River basins defined by the *Eswama,* where the lineages are located.

These conceptual differences are a reflection of historical events and processes that have framed each actor's view of the others. Accordingly, management plans currently proposed for the SNSM necessarily involve the consideration of differing and often contradictory notions of territory, nature, development and cultural difference (participation, agreement, equity and autonomy), and for this reason it is important to analyze how and why these notions differ.

Perspectives on Territory

For indigenous peoples, territory is where cultural identity resides and where the future of their people is written. *Mamas* can interpret within its features the Original Law. Territory is also where natural and spiritual entities live in equilibrium; for them, there is an indissoluble relationship between culture and territory based on the interdependency and mutual determination of the human and nonhuman (see chapter 3).

Indigenous peoples' policies assert their respect for their territory as an expression of the holiness and integrity of human beings' relations to the environment. Territorial management thus implies the conservation and the continuation of indigenous activities in the SNSM.

However, today indigenous peoples of the SNSM see their ancestral territory being fragmented as the result of historical and current land expropriations. This has led to their struggle to recover all their ancestral territory. According to historical circumstances, and in order to negotiate with the state, they divide their territory into: *Resguardo's* territory, *resguardo's* expansion territory, and territory delimited by the Black Line. This latter is the traditional or ancestral territory; the second is the territory that they aspire to recover through the expansion of current *resguardos* and the creation of new *resguardos,*[8] and the first is the legal territory currently acknowledged as *resguardo* by the state. However, from the indigenous peoples' perspective, their territory is one and equivalent to their ancestral territory, and for this reason they have the responsibility of taking care of the whole territory and the lives of both the *elder* and *younger brothers* who live there.

The governmental vision of territory has varied historically and has been fragmented according to the perspective from which it is looked at. The Spaniards' arrival introduced conceptions of possession and property rights necessary to the process of expropriation and colonization by which settlers and peasants displaced indigenous peoples from their lands. From a contemporary governmental perspective, the only indigenous peoples' territories exempt legally from western assumptions about property are the indigenous peoples' *resguardos.* In contrast, a more symbolic rather than practical recognition of indigenous values and authority over lost territories occurred in Resolution 835 of 1995 that recognized the sacred sites and established the Black Line. CPC-91 also states that indigenous territories are those that traditionally have been occupied by the indigenous, regardless if these territories have been or have not been legalized as *resguardos.* Even so, traditional territories do not have the same practical political status as the legally recognized *resguardos.*

From a governmental point of view, the territory of the SNSM has been fragmented by: (1) the administrative political vision that divides it into three departments and fourteen municipalities; (2) environmental jurisdictions that allow different institutions to claim authority over the SNSM: the Ministry of Environment, the SINA entities and three Regional Autonomous

Corporations (CorpoGuajira, CorpoCesar and CorpoMag); (3) the two national natural parks (Tayrona and Sierra Nevada de Santa Marta) and the ecoregion of the SNSM; (4) the indigenous *resguardos*. This fragmentation appears in the various management plans for the SNSM that express differing and often contradictory visions according to each of these institutions' mandates, values and interests.

These different visions of territory determine and complicate all efforts to manage human activities in the SNSM. Although indigenous peoples demand an ancestral and a constitutional right to their territories, these claims do not coincide with governmental policies insofar as the national legal system does not recognize ownership based on ancestral territory. Consequently, these differing visions complicate negotiations. For indigenous peoples, territory is a unity, and its management has to accord to the interpretations of the *Mamas*. As for the government, although it conceives the ecoregion at the environmental level as a single territory, the SNSM is in fact divided into a large number of territorial units in which a great number and diversity of agents make decisions and come to agreements. Nonetheless, indigenous people argue that their traditional management is capable of including such a variety of agreements and agents, and they consider all of them integral and necessary to their view of the negotiating process (see chapter 3).

Perceptions of Nature

Indigenous peoples of the SNSM do not separate the spiritual plane from the natural one or divide nature (environmental issues) from human activities. Humans and nonhumans interact as parts of a whole culture in which the human management of nature emerges from daily actions and practices determined by traditions. They believe there has to be a harmonic balance between the natural and the spiritual based on an energy exchange that is channeled through the *Mamas* whose interpretations of the territory's signs and communication with spiritual entities define individual and collective practices that lead to this balance. These may then be presented as "management" practices as understood by western logic.

In contrast, western science understands the "natural world" (now "the environment") through an objective examination, in which the elements in the environment are alien to the researcher: the natural is "other" to the scientist who studies it. Despite the fact that western science has changed some conceptions of its dualistic view of nature through unifying disciplines such as ecology, in practice it still makes a division of environmental issues into parts or elements for each of which there is a discipline and specific experts. According to Escobar (1999) the term "environment" also "reflects a vision of nature as a resource . . . [In which] what flows is not life, but raw materials, industrial products, pollution agents, resources." Nature is further divided under western logic according to its various roles in other discourses and disciplines such as economics and politics, which fragment even more the relationship between the natural and human. Consequently, agreements between indigenous peoples and governmental and nongovernmental agents require thinking of ecosystemic processes and territorial ordering outside of the modern western framework.

A western vision of modules or parts is especially evident in the scientific and economic approaches to environmental policies. These approaches divide the environment and the territory into scientific and market segments despite an ostensible recognition of the environment as an interdependent ecological whole. Such fragmentation generates separate policy categories such as "conservation" and "indigenous peoples' cultural practices" which, from the indigenous perspective, should be elements of an integrated policy. Consequently, this compartmentalized modern vision is a significant obstacle to the acceptance of cultural practices alien to western scientific and economic logics as valid means to conserve ecosystems and resources. Historically, nonindigenous peoples have used this divide and conquer approach to gain control of indigenous territories.

As a result of this dominant western perspective, national governmental policies value resources and environmental management according to western development models that emphasize the production of eco-products or eco-services for local or foreign markets. In the case of indigenous cultures of the SNSM, such schemes contradict the principle of the right to self-determination and alternative modes of development. The production of eco-products and services does so by placing environmental issues into western scientific, economic and developmental categories that reflect the separation of conservation and economic practices from cultural practices. This vision of the environment as an economic value contradicts the indigenous logic in which the use of resources is an integral element of a simultaneously natural and spiritual context in which there is no purely economic criterion of value.

The Concept of Biodiversity

Biodiversity[9] is another term that has emerged from the discourse of the global environment.[10] Biological diversity is one of the main concerns of conservation practices, and its loss is a global environmental threat. Environmental policies set out three main approaches to biodiversity: conservation, knowledge and use. This division reflects the modern view of nature and the western conservation schemes that have been in use since the 19th century in the colonies. They can be seen in the objectives of the CBD: "the conservation of biological diversity, the sustainable use of its components and the fair and equitable sharing of the benefits arising out of the utilization of genetic resources, including by appropriate access to genetic resources and by appropriate transfer of relevant technologies, taking into account all rights over those resources and to technologies, and by appropriate funding." As global policy, these objectives have yet to be incorporated into the great variety of national and local governmental sectors, plans, programs and policies (CBD 1994). Within the CDB, there are two specific articles (8j[11] and 10c[12]) related to indigenous peoples' knowledge. However, these focus on the need to conserve biological diversity, not the rights or roles of indigenous peoples in enacting such a conservation goal.

These policies demonstrate how nature and indigenous peoples' knowledges have already been subsumed (discursively, if not yet practically) under a market perspective of "resources" and its logic of economic valuation. The recognition of this subsumption is preliminary to the analysis of issues such as genetic material manipulation (Protocol of Cartagena on Biosecurity, 1995) and intellectual property rights in genetic resources (Commission of Cartagena's Agreement, Decision 391 dated July 2, 1996 on Common Regimen on Genetic Resources Access).

In spite of the fact that global environmental policies propose to protect indigenous knowledges and cultural practices, their primary motive is to benefit nonindigenous peoples according to a conservation logic of use and economic development derived from western perspectives.

The Idea of Sustainable Development

The Colombian national government has based its environmental policies on the western concept of sustainable development, and this concept contains important elements that require definition and analysis. The first one is the idea of responsibility for future generations. This idea coincides with indigenous peoples' proposals and practices based on the concept of territorial unity as a means to guarantee environmental balance and to protect life as they know it for future generations. However, the western idea of sustainable development for future generations ignores or minimizes a fundamental aspect of indigenous peoples' beliefs: the past that is the foundation for the future. The ancestral past is the basis of their rights, knowledge, identity, culture and tradition, and it is their basis for legitimizing their land claims and territorial management. However, the west legitimizes its plans according to a hypothetical future in which indigenous peoples and their history have a place only insofar as they do not impede western environmental (or economic) interests. Moreover, the West's future insists upon the creative destruction of the past to make way for the improved future, a concept that portends little tolerance for the value the indigenous place on the maintenance or gradual modification of traditional ways.

The second idea that arises from the emphasis on sustainable development is that the benefits of environmental management belong to humanity. However, the countries and local inhabitants with the greatest biological diversity and the most complex relations with environmentally sensitive areas must bear the cost of protecting this biodiversity for humanity. This implies a collective right—of humanity—to enjoy benefits from the biological diversity that only a few, often poor nations, possess. Not only are environmental resources such as air and water involved, but also access to genetic resources whose use has benefits on a global scale. This vision of biological diversity and development focuses on resources from an exclusively western scientific and economic point of view that largely ignores nonwestern cultural conceptions. The western interest in biodiversity thus puts itself on a collision course with indigenous peoples' collective intellectual property rights.

The term "development" does not exist as such in indigenous peoples' logic, although it has been used in several of their discourses and documents. In the CTC's document for politics in the SNSM (CTC 1999), the indigenous' authorities expressed:

> We understand self development as the exercise of our autonomy; this means that the Sierra constitutes not only the space in which we are developed as indigenous peoples, but also as the only possibility of exercising autonomy [. . .] since the territory of the Sierra is a code in which our authorities interpret the laws that rule nature, that allow our authorities to understand the world's order, and these laws are precisely the ones that identify us as indigenous peoples and provide us autonomy [. . .] development is conceived as a law and standard of enforcement delivered from the origin by the spiritual mother.

In spite of having adopted the term "development" in some documents, currently there is a trend to eliminate it from indigenous discourses and negotiations with the national government because it represents the nonindigenous objectives of national policies with which they do not agree.

Recognition of Cultural Differences: Participation, Agreement, Equity and Autonomy

Despite the fact that national policies, in general, and environmental policies, in particular, recognize diverse knowledges, practices and cultures, there are few cases in which this recognition has been accomplished and incorporated at the practical level in environmental management.

Participation and agreement, despite being significant elements of all current environmental policies, in most cases have continued to be adjusted to western participation and agreement schemes, making the processes ineffective as a means of incorporating indigenous perspectives. The government is currently trying to find approaches to participation that really assimilate indigenous peoples' conceptual schemes, but there is no requirement that the government adjust the western terms of participation and negotiation to accommodate indigenous peoples. This failure to adjust policy-making and political processes to indigenous concepts and objectives effectively limits participation and agreement from indigenous peoples. Attempts to change this situation have resulted in projects such as the ones financed by the GEF, which allocates a period previous to the project's development to permit joint planning with (or cooptation of) the indigenous peoples. However, even this strategy has not secured full participation and agreement because environmental proposals continue to be set out conceptually within western terms and logic alien to indigenous peoples' cultures, particularly the idea of sustainable development. All these situations produce, as a result, a lack of indigenous autonomy and equality in these processes.

These conceptual differences regarding territory, environment, biodiversity, sustainable development and participation generate different approaches to identifying and responding to environmental and political problems. However, there are also other interests, such as those of the peasants, economic sectors and urban populations, which have to be considered. Nevertheless, here the focus is on the contradictions between indigenous peoples and the government (see table 4.3) that arise from how the indigenous peoples define conservation in a manner that combines natural and spiritual elements. This view is schematically expressed in table 4.4.

As explained before, for indigenous peoples, the Original Law organizes their actions with nature, and it is the basis for practices and relation-

Table 4.3. Indigenous Peoples and Governmental Perspectives Related to Environmental Situations of the SNSM

ENVIRONMENTAL SITUATIONS	GOVERNMENTAL PERSPECTIVE	INDIGENOUS PEOPLES' PERSPECTIVES
Identification of environmental problems	Water shortage Loss of biodiversity Deforestation State's lack of -coordination Lack of conceptual unity among entities	Loss of traditional territory in the lower parts of the SNSM Loss and desecration of sacred places Treasure hunting Pollution from deforestation in lowlands of the SNSM and construction of development projects
Concepts on which environmental management is based	Sustainable development. Ecoregion The SNSM as biosphere reserve; important for its biological diversity and water generation Conservation Sustainable use Biodiversity Life Quality Clean production Green Markets Participation and agreement	Original Law that resides in the territory condenses spiritual and material planes, guarantees permanence and balance of territory. Along with *Serankwa* and *Seynekun* it holds the territorial ordering vision *Eswama* where the Original Law is translated and materialized. It is the collective principle of territorial use, management and care *Mamas* who interpret and materialize the Original Law and organize and administer authority The Black Line The SNSM as the world's heart and indigenous peoples as its keepers
Proposals and elements of SNSM's conservation	Ecosystem conservation Strengthening indigenous peoples' identities Peasants' stabilization	Territorial unity and its management and control Promote self-

(*continued*)

Table 4.3. (*continued*)

ENVIRONMENTAL SITUATIONS	GOVERNMENTAL PERSPECTIVE	INDIGENOUS PEOPLES' PERSPECTIVES
	Strengthening human rights Administrative modernization Education for sustainable development	determination and grassroots' strengthening Make progress in recovering, consolidating and controlling traditional territory and sacred places. Consolidation of indigenous peoples' rights. Strengthening of economic, political, cultural, administrative and jurisdictional indigenous practices

ships with their territory. For these reasons, indigenous peoples desire that it form the basis of the SNSM's environmental management. For them, this is predominantly a cultural response to their environment rather than a political or technical one. For the government and different institutions, environmental problems require a political response to technical difficulties identified as consequences of social, economic, scientific, political and administrative processes. Therefore, although environmental management proposals may address indigenous culture, the main emphasis falls upon the technical aspects of realizing proposals for economic development (reforestation, nurseries, agricultural techniques, ecology and ordering of water basins) and their associated administrative components (institutional coordination, strengthening of community associations, strengthening peasants' grassroots organizations).

The western bias of these strategies is evident in the PDS (1997), the official document resulting from four years of consultation and negotiation with all social actors and that identifies different categories of environmental problems. However, its main objective is to "conserve, protect and restore the natural basis of the SNSM to guarantee cultural survival, as well as that of ecosystems and water sources needed for the region's sustainable development." Comparing the PDS with indigenous peoples' policies, there are

Table 4.4. Environmental Problems and Their Effects in the SNSM According to Indigenous Peoples' Perspectives

SITUATIONS THAT GENERATE PROBLEMS		GENERAL EFFECTS	PARTICULAR EFFECTS
TERRITORIAL	Loss of traditional territory on low part of the SNSM	Displacement to higher parts; spiritual imbalance; breaks in social order by basins and lineage	Lack of lands for traditional practices, over population in specific places, natural/cultural imbalance
	Loss and desecration of sacred places	Weakening of the Mother; spiritual imbalance	Disease increases, deforestation, Breaking nature's balance
EXTERNAL ACTORS	Treasure hunting	Looted sacred places that prevents *pagamentos* and ritual practices	Imbalance with - the Mother
	Pollution from deforestation in low parts of the SNSM and construction of development projects	Spiritual imbalance affecting sacred places	Reduction in river, lake and swamp levels during summer, reduction of fish
	Governmental institutions without coordination	Disrupting authority and traditional laws	Introduction of national and global environmental policies that interfere with local autonomy

several significant differences. One of the principal differences is the idea of sustainable development that frames it, which strongly contrasts with indigenous peoples' view that cultural survival will guarantee the ecosystem's conservation (see chapter 3).

The PDS sets out six different goals. One of these is related to strengthening the indigenous peoples' cultural identity. It identifies courses of actions needed to "solve" indigenous peoples' problems, and in some of these the PDS coincides with several of the indigenous peoples' proposals. For example, the PDS's plans promote respect for indigenous territories, and more specifically, their sacred places, thereby strengthening their culture, traditional environmental management techniques and the process of recovering their territories.

Despite these coincidences, the PDS's approach to ecosystem conservation specifically excludes indigenous peoples' cultural (ecological) practices, such as the *pagamentos* in sacred places, among others (see table 7.5). This demonstrates that regardless of the conceptual recognition that environmental policies give indigenous peoples and their environmental knowledge and practices the government does not recognize indigenous peoples' practices as viable conservation strategies in the field.

For indigenous peoples, environmental policies and practices must not be derived from abstract conceptions since the territory has written in its features the Original Law that determines appropriate environmental practices. The *Mamas* interpret the territorial laws and determine actions that will guarantee harmony between the material and the spiritual planes for both humans and nonhumans. It is for this reason that they refuse to prepare rigidly-planned, and specific projects, since these do not fit with their more process-oriented and situational manner of thought. These sorts of problems and conflicts are also evident in the national environmental research policy (2001), which does not sufficiently consider nonwestern approaches to economic resources and imposes a western logic of development.

There are also instances when indigenous peoples' policies contradict their own traditional practices and locate their claims within the western perspectives of state discourses. Although autonomy to manage natural resources according to their practices is one of their principal demands, they nonetheless accept the possibility of western conservation or development schemes in territories that are to be enlarged or likely to be attached to their *resguardos*. This is evident in the policy draft document prepared on October 2001 by the CTC that states that these areas (that will be enlarged *resguardos*) could be "developed" through "inter-institutional, intercultural and complementary cooperation upon different zones according to the conservation approach (forest reserve, buffer areas combined with *resguardo* expansion zones)." The indigenous use of the state's environmental plans in certain ways implies better territorial recognition and protection; but implicitly it gives recognition to the priority of the state's conservation strategies over those of the indigenous peoples.

In other cases, indigenous peoples have proposed and implemented national environmental policies that coincide with their values. For example, conserving the SNSM as a human heritage and a biosphere reserve reflects their belief that their management of natural resources promotes *biodiversity and water* conservation in ways that contribute to the global well-being of *humanity*. And insofar as water and biodiversity are fundamental elements for national and global environmental policies, indigenous peoples receive state and NGO support for developing these goals.

Indigenous peoples often use conservation arguments pertaining to water as a political strategy to consolidate their proposals within the national and global environmental contexts. However, these arguments do not necessarily contradict their traditional values because water is also a vital element for the indigenous peoples' way of life, as both a practical resource and symbolic element. Indigenous peoples' settling practices are related to water basins, and the ecological management of those habitats depends on the spiritual and environmental interrelations and specificities of each basin. Indigenous peoples base their social organization on lineage distribution, which they determine by basins, and each linage has specific responsibilities in relation to the basins that they occupy. They consider water a female element associated with fertility and life. Lakes are sacred places and the sea is the Mother. Accordingly, the assertion that SNSM conservation would benefit humanity coincides with the belief of the indigenous peoples that the SNSM is the Heart of the World and that from its maintenance the rest of the body benefits, in this case the rest of the world-universe. In other words, they understand that caring for the SNSM is a duty and right granted ancestrally, and that the rest of humanity obtains benefits from their stewardship.

In a document from the Territorial Cabildo Council (February, 2002), indigenous peoples of the SNSM established other environmental management actions framed in western terms. Among these actions are: defining protected zones to regenerate the forests and as a recovery and conservation plan for river basins and micro-basins. Likewise, they propose to continue with their practices, such as cultivation rotation, diversified agriculture and use of diverse seeds as means to improve the environmental condition of the SNSM (see table 7.5).

The contradictions in indigenous peoples' thinking and activities denote indigenous peoples' struggles to achieve political strategies that legitimize their discourse in western contexts. In this way, the *ecological native* is also the product of the indigenous peoples' movements' politics as well as nonindigenous national and global policies. The discursive conflicts evident in indigenous thought and action results from the various environmental agents and interests that generate a constant interaction, contradiction and negotiation within global environmental processes that reconfigure the practices and identities of all those involved in them.

FINAL REMARKS: CONFRONTING THE GLOBAL ENVIRONMENTAL ARENA

Ecological awareness concerning the SNSM is related to the deterioration of different ecosystems and the shortage or extinction of natural resources that has transformed not only ecological processes but also cultural practices, both western and indigenous, and the intricate symbolic networks founded upon these resources. In the same manner, this awareness implies new forms of interdependence and global control of environmental problems that involves national government, grassroots organizations, nongovernmental organizations, experts and local people.

An emergent global eco-governmentality is introducing new ways of framing and directing local knowledges, participation, and territories in terms of modern western concepts. Correspondingly, indigenous peoples often have to frame their contributions in the same logic of management that has been used since the colonial periods. As we saw in the 19th century, these consisted of developmental environmental policies implemented by the colonial empires, or processes developed in Africa through the implementation of agricultural techniques described by Comaroff and Comaroff (1997). In a similar way, an eco-governmentality can restrict the indigenous peoples' ecological identities through western disciplinary mechanisms that impose a particular relation to the environment: principally that of progressive development.

The modern western view has created various representations of indigenous peoples that focus upon their ecological roles and their territories (as we will see in the following chapters). These representations generally assume that indigenous peoples live according to a conservationist ethic, but they also assume that that conservationist ethic will be consistent with the interests and practices evident in western logic and global environmental policies. Such a global eco-governmentality implies conserving priority areas (hot spots and biodiverse areas) from a biological point of view, especially in areas owned ancestrally by indigenous peoples; and it assigns to these peoples the responsibility of maintaining "humanity's heritage." In the western process of disciplining green-bodies and minds, indigenous peoples, through processes of mapping, planning and the implementation of environmental strategies, have been integrated, willingly or unwillingly, into new relationships with their territories and resources.

Nevertheless, indigenous peoples find in these western images and environmental policies opportunities for more and better recognition within national and transnational environmental discourses. They therefore consider this historical situation as an opportunity to consolidate their historical struggle to defend and recover their territories and as a way to be autonomous according to their own traditional life plans. In other words, this can be an opportunity for indigenous peoples to construct ecological identities based on their own knowledges and practices.

Indigenous peoples' documents and positions related to the environment express concerns that transcend western environmental discourses that emphasize development. For example, the *Mamas'* decision made in a meeting in September 2001 states that they do not want to participate as beneficiaries of the development of "economic resources" planned by the PDS, thus indicating that the PDS's projects will not be carried out in *resguardo* territories or in zones provided for their expansion because the *Mamas* do not agree with the western notion of development.

In the SNSM, indigenous peoples' main proposals have been permanently documented during more than five years of negotiation with the state in relation to the SNSM. In the Draft Policies Documents of the CTC (November 1999:19), they state:

> Our consideration of any development plan is precisely to promote indigenous peoples' environmental views as a fundamental core to make out of the Sierra the territory that we all want: one that is conserving and producing life. Thus, autonomy is not one of the tasks, but the main one. Indigenous peoples cannot be separated from any other social actor [or process] as conceived by the *younger brothers*. Autonomy is the driving axle of any deep transformation that will be promoted to save the Sierra.

Such a statement indicates how the creation of a global environmental discourse will involve a constant process of struggle over environmental meanings. In it we also see how indigenous peoples are challenging the underlying western assumptions of such global strategies and proposing alternatives to them.

The contradictions and syntheses of these different conceptions all participate in the formation of an environmental discourse of global eco-governmentality. This discourse is not yet hegemonic in character because it still offers the possibility of contesting, challenging or transforming each participant's unique vision of environmental problems and solutions. Moreover, the effort to make it a discourse of life (human and nonhuman) promises to open new issues that will pose further challenges to all people involved. This will occur as new green networks create new relationships that reconfigure previously established systems and discourses of power/knowledge. In fact, various political strategies and actors have already made significant advances in promoting such changes: nature-cultures (indigenous peoples), philosophical actions (Deep Ecology), political activism (Greenpeace or suburban zoning commissions), eco-subversion (Earth First!), eco-capitalist entrepreneurs (The Nature Conservancy), technobiologism (Biosphere) and local knowledges (those of the indigenous peoples of the SNSM).

This is not to deny that among these diverse agents and interests there are some that have more economic and political power to impose their strategies and act as if they have become hegemonic. Such power is evident in the case of Colombia where a neoliberal version of eco-governmentality has rather successfully imposed itself in order to gain control of the nation's economic relations and natural and genetic resources. However, the efforts of indigenous peoples and the many other environmental actors to bring alternative environmental visions to the debate over global eco-governmentality indicate that such a hegemony is unlikely to become a matter of fact.

Chapter Five

Ecological Identities of Indigenous Peoples: Historical Process of Construction

INTRODUCTION

In this chapter I explore the historical circumstances and processes of the construction of indigenous peoples' ecological identities. This analysis is of a two-fold nature: on the one hand it requires consideration of how and why western peoples constructed representations of indigenous peoples and subsequently integrated them to western environmental movements and discourses; and on the other, it requires consideration of how indigenous peoples have responded to, appropriated or negated such western processes, representations and discourses, especially with respect to indigenous ecological identity. The ultimate aim is to address how this historical process of representation and integration has affected indigenous peoples' social movements and collective identities in Colombia, in general, and those of the indigenous peoples of the SNSM, in particular.

Identity construction theories (Hall 1990, Scott 1995, Wade 1997, Castells 1997, Comaroff and Comaroff 1997) are particularly useful for the analysis of indigenous ecological identities because of the intricate links they identify between the cultural politics of indigenous peoples' movements and global environmental policies and politics. Of particular importance for this analysis are the links between tradition (territoriality, historical memory and customary daily activities) and new political strategies (new forms of political organization and identity construction). This is so because indigenous peoples use their past in both its historical and traditional modes as the primary means to maintain or modify the historical continuity of their identities and produce political power.

Ecological identities require analysis as the products of often contradictory discourses generated by various agents situated at different points on the spectrum of power, including indigenous peoples' cultural politics, transnational entities, expert knowledge, ENGOs, indigenous peoples' organizations, the state, and neoliberal policies. Consequently, I map indigenous peoples' ecological identities through the practices, ideas, representations and discourses of both western and indigenous peoples from local to national and international arenas.

In this chapter, I also discuss how the new articulation of indigenous peoples and the environment not only constructs new identities but also causes issues related to territory and political autonomy and sovereignty to differ from such issues in earlier indigenous peoples' struggles. Although these new indigenous claims regarding autonomy, sovereignty and territorial rights arise and take form within the dominant political order and interests of the nation-state, their environmental dimensions transcend the national context and prompt the nation-state to form new political ends and projects in response to the supranational scope of their claims.

Indigenous peoples have thus promoted new national and international conceptions of natural resources and access to genetic resources that take into account indigenous demands for self-determination and sovereignty that require the recognition of their own cultural and political authority in their territories. I address here how the environmental politics of indigenous peoples' movements articulate their local territorial demands to a process of identity construction with national and global implications.

IDENTITY POLITICS

Since the end of the 1970s, indigenous peoples' political actions and processes of identity construction have been related to ecology, environmentalism and conservation (Conklin and Graham 1995, Varese 1996b, Conklin 1997, 2002, Ramos 1998, Brosius 1999, 2000, Bengoa 2000, Brysk 2000, Ulloa 2001). This identity often takes the form of images and representations that invoke indigenous peoples as stewards of nature, eco-heroes, or *ecological natives* who protect the environment and represent hope in the face of global environmental crisis.

Many of these new (and largely western) representations recognize indigenous peoples as keepers of ancestral knowledge that allows the continuity of biological diversity and harmony within their territories. Consequently, this identity promotes new western stereotypes of indigenous groups and local communities in which they appear as wisely innocent inhabitants of an ecological paradise (preferably a tropical rainforest). Such stereotypes of *ecological natives* are, for the most part, independent of indigenous peoples' traditional systems of knowledge and current political goals. They arise instead from representations in the western mass media, environmentalist rhetoric, and romantic idealism regarding a lost paradise that in fact are expressions of western environmental desire, concern, fantasy and guilt (see chapter 6).

Some anthropologists and historians have examined the relationship between indigenous peoples and the environment to determine if there ever was a harmonious relationship between them and the environment prior to colonial encounters. These scholars claim that such a harmonious relationship was indeed possible before the conquest, but it was broken by the introduction of commercial circuits of fur, feathers, pets and meat. In order to explain this transformation, they use several approaches: indigenous peoples were seduced through desire for new western commodities, although they did not completely lose their worldview; indigenous peoples had a spiritual crisis because of the encounter and the harmony of their relationship with nature suffered; and, finally, indigenous people have in fact maintained their harmonious relationship with nature because it has been a constant element of their resistance strategies throughout the historical process of contact with the West (Varese 1996b, Luca 2001).

Varese (1996b) points out that the relationships of indigenous peoples with their environment are responses to a complex historical process of interrelation between cultural and ecological practices. He states that western contact reinforced these relationships as political strategies of resistance in four ways: increased commitment to the management of the universe through an ecological ethic; emphasis on economic relationships based on reciprocity; more secrecy regarding the biological knowledges that allow the indigenous to both exploit and maintain biological diversity; and, finally, the remarkable political flexibility and plasticity of the traditional discourses of indigenous peoples.

Indigenous environmental identity has become an interplay of power relationships that resembles the construction of ethnic identity that occurs between colonizers and colonized in which rebellion and resistance arise as immediate effects of attempted conquest (Bonfil Batalla 1981). "Ethnic consciousness," as Comaroff and Comaroff (1997) point out, "has its origins in encounters between peoples who signify their differences and inequalities—in power, economic position, political ambitions, and historical imaginings—by cultural means. Typically, it is the subordinate, not the dominant, who is first marked and named" (388). And so it is that indigenous peoples have had to respond to the west as *ecological natives*. Nonetheless, because this is a "dialectical encounter," collective identity is an historical process that can be changed because it is "everywhere a relation, nowhere a thing" (Comaroff and Comaroff 1997: 388). Identity is not essential, and thus in their terms:

> [T]he *content* of any ethnic identity is a product of complex, drawn out historical processes: being a heterogeneous, fluid ensemble of signs and practices, a living culture is forged not merely in conversations, but also in the minutiae of everyday action, in the inscription of linguistic forms and material relations, in the course of struggle, contestation, and creative self assertion. (1997:389)

Accordingly, the relationship between the western environmental and indigenous peoples' movements formed the bases for the consolidation of indigenous ecological identities during the 1970s. Since then, there has been a "natural alliance" between the environmentalists and indigenous peoples which has been confirmed through the national and international conventions related to the environment. Such recognition is also a more visible effect of the political struggles of indigenous peoples to assert themselves and demonstrate how their own cultural and environmental politics articulate not just a strategy of resistance but a proactive plan addressing both indigenous and nonindigenous issues concerning identity, cultural practices, territory and the environment.

In 1992, during the quincentennial of the discovery of America, indigenous peoples asserted their political processes, their claims for autonomy, and their rights to manage their natural resources and territories according to cultural practices and conceptions of nature that differ from modern western notions. The quincentennial thus gave global visibility to the relationship between indigenous peoples and environmentalism. In this manner, indigenous peoples linked their cultural politics and processes of identity construction to national and transnational environmental projects. As Alvarez, Dagnino and Escobar (1998) argue, such strategies have allowed indigenous social movements to gain the political power they need to rename and transform the concepts of nation, citizenship, development and democracy that define modernity itself. In short, by accepting the environmental designation, indigenous peoples' movements have given themselves an opportunity to rethink basic western categories of thought, especially nature.

Identity construction, as a general idea, is related to self-differentiation. As Barth (1969) pointed out long ago, the self-identification of actors within specific ethnic groups is basic to the ascription of ethnic identity. So, too, the idea of self-differentiation is basic to the process of identity construction within social movements. However, the idea of self-differentiation is also a relational process that responds to identities that have been historically conferred by others. Hall (1990), analyzing the African Diaspora, stresses that cultural identities and their transformations are in a dialogic relation with the past (continuity) and the present (discontinuity and rupture). Identity construction is thus a negotiation of history, power and culture in the specific locations in which it takes place or, as Wade notes (1997), identity also has to be analyzed in relation to class, sexuality, gender, race and religion.

In a similar way, Scott (1995) stresses the necessity of historicizing the question of identity. Her "strategy . . . is to introduce an analysis of its production, and thus an analysis of constructions of and conflicts about power; it is also, of course, to call into question the autonomy and stability of any particular identity as it claims to define and interpret a subject's existence" (8). Consequently, identity has to be analyzed as the relational result of repeated processes of cultural interaction rather than the emergence of historically categorical entities. She argues that identity is "the unstable, never-secured effect of processes of enunciation of cultural differences" (1995:11). Ethnic identity can thus be considered an historical construction or process that is at the same time conferred, assumed and challenged by all involved in its construction.

In spite of diverse orientations to the problem of identity, these scholars agree that identity construction is a process of self-differentiation (continuity) in relation and opposition to conferred identities that takes place within specific historical processes and relations of power/knowledge that imply negotiations and conflicts. In addition they agree that historical encounters between "self" and "other" have contributed to the construction of specific categorical identities or stereotypes of the other (antimodern, primitive, or undeveloped) and the dominant self (civilized, tolerant with difference, or multicultural).

Nevertheless, categorical identities have also helped indigenous peoples to reconstruct and invent differences as a strategy that challenges their social locations in the dominant western order. Participation in the western processes of identity construction may allow nonwestern individuals to become political subjects whose agency can transcend the limits of a categorical identity ascribed to them by transforming it into a collective identity that enables and coordinates their political actions. Therefore, a categorical identity can be contested, transformed, reinterpreted or, maybe, abolished by those to whom it is ascribed. Furthermore, collective identity can transcend the collectivity to form itself insofar as it is a relational response to multiple agents and situations (nonindigenous in this case) that require it to redefine itself constantly as a new social "entity". From this perspective, the *ecological native* is not only a stereotype, but also a useful and effective means of indigenous self-representation in nonindigenous arenas.

Indigenous peoples are in constant interaction and negotiation with global environmental processes that reshape both local and transnational practices, representations, discourses and identities. In this interrelation different actors (the state, multilateral institutions, transnational corporations, environmental nongovernmental organizations, social movements, expert knowledge, local actors and researchers) construct new conceptions of nature and environment which, in turn, the cultural politics of the indigenous movements constantly challenge and reshape. Consequently, a specifically ecological indigenous identity has arisen from the intricate links between their cultural politics and nonindigenous environmental policies and politics at the local, national and international levels.

The indigenous peoples' understanding of their own collective identity as a result of historical processes and contemporary social, political, economic and cultural dynamics is far from any stable or unitary holism presupposed by western discourses of primitivism. This understanding has allowed indigenous peoples to construct flexible political strategies articulated to ecological practices in a manner that promotes respect for their ethnic identity and enables them to respond to the ever-changing political relationships of culture and territory. In Latin America, the indigenous struggle to assert identity and difference in relation to territorial claims has continued for more than 500 years. However, indigenous peoples' struggles do have distinctive articulations in specific historical moments. Of the many noteworthy instances of indigenous resistance to nonindigenous agents and agencies during these periods, I will focus only on those that we would now consider relevant to the topic of the environment.

In the Colonial period, the indigenous peoples' struggles to defend their sacred conceptions, places and rituals resulted in some cases from a conflict between two different biotas made evident by the introduction of European plants and animals (Crosby 1988). During colonial times, several uprisings led by Mayan, Totonac, and Quechua peoples were "motivated by a call to reconstitute the social and cultural precolonial order and return to the world's sacredness; to purify a nature contaminated by foreign oppressors"

(Varese 1996b:125). In 1546, in Mexico, the Mayan peoples rebelled against the Spaniards; the objective of this revolt was to purify the land and to end the Spanish domination. During this insurrection, the Spaniards were sacrificed along with their plants and animals. In 1780, the Totonac vindicated their revolt against the Spaniards with reference to their trees: "The trees give shade to people and help them to persevere, are useful to tie animals, protect houses from fires, and the branches and leaves are used as fodder for animals" (Taylor 1979:137 quoted by Varese 1996b:125). In the Amazon region of Peru the Quechua and other ethnic groups rebelled against the Spaniards and Peruvians in 1742 and made the following demands: "the rights to live in dispersed villages and households; to allow a rational use of the tropical rain forest; the eradication of European pigs, which were considered harmful to farming and peoples' health; the right to free cultivation and the use of coca, 'the herb of God,' and the right to produce and ceremonially drink masato, the fermented manioc beverage of substantial nutritional value" (Varese 1996b:125).

During the Republican period in Colombia (the latter half of the 19th and first half of the 20th centuries), the indigenous peoples struggled against the politics of a central government that proposed to integrate them and their territories under new, modern ideas of production and political participation (a process that continued through the first half of the 20th century). Under the logic of the new republic "citizens," including indigenous peoples, were "equal." Indigenous peoples' differences were thus suppressed as part of the project of nation building that promoted a political and economic individualism antithetical to indigenous collectivism. This suppression is also evident in the tendency of 19th and early 20th century western history to generally diminish the importance of indigenous peoples' resistance, even though they had a constant and important effect on that history, especially in Latin America. Nevertheless, indigenous peoples continued their struggles to defend their territories, to retain access to fertile land and to assert collective property rights over their land in response to the modernizing process that occurred during the republican period.

Finally, a new period centered on the protection of the environment arose in the latter half of the 20th century that could be called the "environmental era." In this period, indigenous peoples' claims increasingly focused on the political autonomy they needed to control and manage the natural and genetic resources in their territories. This change occurred in large part due to the new economic value of nature in a global economic and political order that again challenged their cultures, identities and claims to local political sovereignty. In this period (which continues to this day), claims related to territory differ because indigenous political autonomy and collective property rights affect, and have been affected by, global environmental and economic agendas that involve conservation and industrial development (biospheres, parks, oil exploration, and pharmaceutical research, among others). For these reasons indigenous peoples have argued for more democratic (indigenous) participation in matters of resource management. Accordingly, during the last two decades, their political claims have focused on environmental interests, thereby influencing and changing their own ideas of nature, territory, autonomy and identity.

TRANSNATIONAL RELATIONSHIP BETWEEN INDIGENOUS PEOPLES AND ENVIRONMENTALISM

Indigenous peoples' efforts to link their collective identity to concepts such as territory, political self-determination and autonomy have allowed them to establish a clear connection between culture, territory and environment. At the same time, the discussions of the environmental movements during the 1970s and their proliferation in the 1980s also promoted the integration of indigenous peoples' movements with western conceptions of the environment (Bengoa 2000).

Indigenous peoples acknowledge that encounters and coalitions have fostered initiatives that recognize and respect the rights and vital needs of indigenous peoples as integral to the need to preserve biological diversity. The consolidation of this indigenous/environmental coalition mainly took place during the decade of the 1990s. As a result, environmentalists and indigenous peoples have built new perspectives (COICA 2001). The coalition between the indigenous and environmentalists has often been productive for both, as the Coordinadora de las Organizaciones Indígenas de la Cuenca Amazónica (COICA) states:

> These alliances have given indigenous peoples the possibility of a wider and more active participation in debates and discussions in the international and United Nations agendas. Therefore, the various alliances are a valid and efficient alternative to globalized struggles, resistances and solidarities of indigenous peoples. Also, these alliances have helped to reaffirm technical and financial cooperation with indigenous peoples. (COICA 2001:57)

In turn, environmentalists and conservationists have benefited from the authority indigenous peoples give to their interests. In countries such as

Ecuador, Brazil, Mexico, Nicaragua and Colombia, indigenous peoples and environmentalists have struggled together against oil and lumber companies, development programs, hydroelectric and highway developments, and bioprospecting research.

A Short History of the Coalition in International Political Arenas

In 1971 the Man and the Biosphere (MAB) program of the United Nations Educational, Scientific and Cultural Organization-UNESCO began the scientific investigation and gathering of information related to traditional knowledge of the use of natural resources. In this manner, MAB promoted a close relationship between the environmental knowledges of local ethnic and indigenous communities with the western interest in the conservation and sustainable use of natural resources. In 1972 the Convention Concerning the Protection of the World Cultural and Natural Heritage confirmed the relationship between indigenous peoples and the environment when it began to integrate actions both to protect cultural identities and conserve nature. This could be considered the first "official" international recognition of the relationship between indigenous peoples and specifically "environmental" concerns.

In 1980, The World Conservation Union-IUCN, through the World Conservation Strategy, highlighted the relationship between ethnic groups and the conservation of nature, ratifying a correlation between indigenous cultures' practices and the conservation of ecosystems of importance for biological diversity. Likewise, it stated that the management of protected areas should also promote economic development of the local residents, which created further links between conservation and local inhabitants. In a similar way, during the 1980s, due to the scientific concern related to the environmental crisis (global warming, deforestation, and pollution), environmentalists linked their concerns with indigenous peoples' demands for the recovery of their territories.

In 1990, in Iquitos Peru, indigenous peoples and environmentalists wrote a declaration that stated that indigenous peoples and their territory are one and that to destroy one is to destroy the other: "We hold that legal recognition and defense of the territorial rights of indigenous peoples is an essential component in the management and conservation of the Amazon" (Declaration of the Alliance 1990). Later, in 1993, in Washington "A Working Conference to Protect the Indigenous Rights" was held. Both of these historic meetings were inspired by COICA participants who argued that the best defense of the Amazon Basin is support of indigenous peoples' claims to territory and the development of western environmental policies and strategies consonant with indigenous cultural practices.

This was the beginning of the coalition between indigenous peoples and representatives of COICA and U.S.-based environmental organizations that formed the Amazon Alliance for Indigenous and Traditional Peoples of the Amazon Basin. This alliance works to defend the territorial and the environmental rights of indigenous peoples and traditional peoples of the Amazon Basin around the following thematic axes: legislation and international agreements, economic strategies and life plans according to traditional cultural practices, natural environment and cultural territories, and human rights.

The general coalition of indigenous and environmental movements received international recognition in 1991 when UNESCO sponsored an indigenous meeting in Paris as a preparation of the World Summit. In 1992, in Santiago de Chile, the United Nations Technical Conference on Indigenous Peoples and the Environment established some of the basic principles that form the basis of the relationship between the environment and indigenous peoples' territories and their right to political self-determination. As Conklin notes:

> In the 1980s and early 1990s, there was widespread optimism that the best way to protect the integrity of environment was to secure native lands rights, thereby keeping the control of environmental resources in the hands of Native peoples who have protected those resources for centuries. (Conklin 2002:1054)

Finally, in 1992, in Rio de Janeiro, the United Nations Conference on Environmental Development (UNCED) consolidated the union of indigenous peoples and environmental discourses as confirmed in the CBD (although not completely in the terms proposed by indigenous peoples), particularly in articles 8j and 10c, and in the Agenda 21. From this moment the leaders of the world recognized openly the relation of indigenous knowledges to the protection and conservation of the environment. Indigenous peoples consider these events very important because, through them, indigenous peoples gained global recognition as environmentalists.

At that time, indigenous leaders also established links with environmentalists, as in the case of Davi Kopenawa, a Yanomami shaman, Raoni, a Kayapó elder chief, and Payakan, another Kayapó leader, all of whom participated in various environmental events that helped situate the indigenous peoples' environmental and territorial discourses within transnational ecopolitics (Conklin 1997).

The articulation of indigenous peoples and the environment ratified in Rio-1992 generated a series of initiatives like those of the Earth Council[1] and the Fetzer Institute[2], which carried out different meetings to reaffirm the bonds between the environment, spirituality and indigenous peoples' rights. These institutions carried out the project "Indigenous Peoples, Mother Earth and Spirituality" through consultations and meetings in Costa Rica (1996), Argentina (1998) and Honduras (1998). These meetings generated declarations, official statements and processes that recognize the knowledge that indigenous peoples possess to manage natural resources in a sustainable way and with due respect for Mother Earth.

Likewise, alliances have been built that allow common purposes between ENGOs and indigenous peoples' organizations. Such is the case of the International Alliance of the Indigenous-Tribal Peoples of the Tropical Forests-IAIP (http://iaip.gn.apc.org) that formed in 1992 in Malaysia, and that was attended by indigenous people and delegates of environmentalist NGOs that represented indigenous peoples and indigenous leaders. Their actions have focused on the rights of indigenous peoples of tropical forests in relation to the environment, natural resources and international law.

In 1993, the World Wildlife Fund-WWF and the IAIP agreed to produce a general policy related to conservation and indigenous peoples. This agreement recapitulates the principles of the current Draft Declaration on the Rights of Indigenous People-UN recognizing territorial rights of indigenous peoples and the necessity of a free, prior and informed consent to initiate any activity between NGOs and indigenous peoples. In a similar way, in 1996, the IUCN approved resolutions concerning indigenous peoples following the Draft Declaration on the Rights of Indigenous People-UN whereby the IUCN recognizes the territorial and collective rights of indigenous peoples (COICA 2001).

The interrelation between indigenous peoples and environmentalists has been only one part of the process of constructing their ecological identities. Another important part are the policies and actions of governmental and transnational organizations evident in the ratification of agreements, conventions, declarations and transnational arenas: the International Indigenous Forum on Biodiversity (1996), the Ad Hoc Open-Ended Inter-sessional Working Group on Article 8 (j) and Related Provisions of the CBD (2000), the Ad-Hoc Open-ended Working Group on Access and Benefit-Sharing (2001), the International Forum of Indigenous Peoples and Local Communities on Climate Change (2000), the Indigenous Peoples' International Summit on Sustainable Development, and the Johannesburg declaration, among others.

As a result of these dynamic interactions, *transnational environmental indigenist movements*[3] (TEIM) have been constructed. The philosophical discourse that sustains this coalition promotes a relationship between humans and nonhumans as a matter of mutual respect, order and balance: "It has been constituted in a symbolic force, true or false it does not matter, about what was and what can be the system of life, in which natural orders, the order of the men (and women) with other men (and women) and the order of these with nature are restored" (Bengoa 2000:75). Ecological identities have thus been constructed through an interplay of local, national and transnational discourses promulgated by indigenous and nonindigenous actors at each level.

The actions of TEIM in support of indigenous peoples movements, particularly in the case of indigenous peoples in Colombia, could be seen as powerful and victorious: a triumphal movement that in the last decade has changed global and national environmental law and transnational nongovernmental and governmental institutions by introducing demands for the recognition of indigenous peoples' collective rights in national and international arenas. However, transnational institutions and the TEIM have had contradictory effects in national and local settings. They often promote specific ideas of *ecological natives* and specific ideas of leaders that are not always representative of indigenous communities, though they may correspond to transnational expectations of an *ecological native* and what is appropriate for his or her discourse. As Tilley (2002) states in relation to Transnational Indigenous Peoples Movements (TIPM), "each institution assumes its own incontestable authority, as a morally-driven actor, to arbitrate that discourse's tenets" (553).

Thus TEIM have constructed their own ideas of the qualities, needs and actions that indigenous peoples must pursue to resolve their problems. Tilley states that "TEPM concepts, as deployed locally, manifested as a newly hegemonic discourse of indigeneity" (Tilley 2002:553). As she shows in the case of Salvadorean indigenous peoples, when local organizations or movements do not conform to these transnational standards, the results are contradictions among or negations of local indigenous identities or, at worst, nonconforming indigenous peoples are simply considered nonindigenous.

TEIM and NGOs often have an idea of indigenous peoples and their practices that assumes a total interrelation between nature and indigenous culture that places the indigenous peoples in the category of nature and the western peoples in that of culture. This dichotomy affirms the dualist western conception of nature and culture. As a result, indigenous peoples' societies become external natural entities under the "law of nature," while western societies are "rational" cultures governed by intellect. And, according to western logic, culture should ascribe value to nature.

These categorical representations imply that local organizations should shape their proposals, projects and requests for financial support according to a nature/culture model. Consequently, these ideas often frame the environmental discourse that introduces indigenous peoples into planning and the elaboration of management programs of natural resources (as we have seen in the previous chapters). For example, the international discourse of Unesco related to culture heritage and biodiversity links indigenous peoples to western ideals of traditional and ecological practices. In the international arena, indigenous peoples are often represented as uniquely ahistorical in their "cultural autonomy, homogeneity and rigidity (visually inscribed in dress and other practices)" (Tilley 2002:553). Such ideas reaffirm and essentialize images of indigenous peoples insofar as TEIM have created multiple bureaucracies to coordinate indigenous peoples' work that receive more financial support the more they can portray the indigenous as "traditional" and "ecological" (part of nature).

However, environmental discourses have diverse characters, and not all coalitions of indigenous peoples and environmentalists employ a rigid nature/culture dichotomy. Inside these coalitions, there are relationships among environmentalists and indigenous peoples that address these diverse interests and that transcend fixed categorical representations, from protecting biodiversity and indigenous territories to promoting eco-tourism, ethno-tourism and bioprospecting.

Although there are many successful alliances between indigenous peoples and environmentalists, there often remains a conceptual divergence between them. According to COICA (2001), indigenous peoples highlight the following causes of the global environmental crisis: the modern social structure, the dominion over nature as a result of the idea of men's dominion over women, and, finally, the hegemonic market-economy and development model. This conceptual difference appears in strategies and actions that the indigenous propose as means to address the environmental crisis.

> For indigenous peoples conservation necessarily goes with the recognition of our territorial rights, the protection of our cultural and intellectual patrimony, the harmonic use of natural resources that allow the survival of our cultures, and the overcoming of our situation of being dominated by others. (COICA 2001:192)

Consequently, indigenous peoples argue that conservation actions led by environmentalists, as in the case of protected areas, have been: 1. Defined without, free, prior and informed consent and without the appropriate participation of indigenous peoples; 2. Controlled and administered by nonindigenous state institutions; 3. Imposed through exogenous conceptualizations and practices that do not recognize indigenous peoples' traditional notions and practices related to sustainability; 4. Implemented as environmental management strategies that do not accomplish the priorities defined by indigenous peoples and which contradict indigenous peoples' proposals for the management of their territories and resources; 5. Implemented without considering that the derived benefits of conservation, current and potential, should be shared equally with indigenous peoples who live in protected areas (COICA 2001).

Nonetheless, indigenous peoples have located their demands and positions within these TEIM. They have thus gained recognition as *ecological natives* from national and international institutions and situated themselves as important agents within transnational ecopolitics. In this sense, they have succeeded in assuring that the interactions between transnational institutions and indigenous peoples' discourses have a mutual rather than unilateral character.

INDIGENOUS PEOPLES OF THE SNSM IN THE GLOBAL ENVIRONMENTAL CONTEXT

In Colombia the relationship of indigenous peoples of the SNSM and environmental movements and discourses has diverse origins. There the interplay among indigenous peoples, ENGOs, TEIM and national environmental policies forms an ideal arena for international and national policies and environmental concerns. In particular, the Kogui people have been introduced to TEIM because they conform to the transnational ideal of *ecological natives* as close to nature and far from western society, which appeals to a western romantic sensibility and the dream of returning to a natural paradise.

The reality is that the peoples of the SNSM have become involved with TEIM for a number of different reasons. In the 1970s, with the discovery of the Lost City (Ciudad Perdida), there was a new period of awareness and positioning of indigenous people of the SNSM as bearers of the "traditional" ecological knowledge expressed in the environmentally-sensitive engineering used to build the terraces in the Lost City. The cultural continuity established between the pre-Columbian inhabitants, the Taironas, and the contemporaneous indigenous groups permitted the claim of a linear historical environmental relationship between those indigenous peoples and their territories. Other factors in Colombia have also helped to consolidate this relationship, such as anthropological discourses, environmental programs, and the effects of ENGOs and indigenous peoples' organizations' policies. Independently of the current ecological support that this version of indigenous identity promotes, I would like to focus on the other historical actors and sources that have encouraged the construction of the ecological identity of the peoples of the SNSM (Uribe 1988, Orrantia 2002).

Many perceive indigenous peoples of the SNSM, specifically the Kogui people, as the "ecological indigenous peoples" par excellence (see chapter 6). This positioning is a result mostly of anthropological analyses, descriptions and discourses that address indigenous peoples and their relationship to environmental concepts of Mother Earth and fertility. The anthropological passages about the Kogui people in the first ethnographies of the SNSM by scholars such as Konrad Preuss, Gregory Mason and Gerardo Reichel-Dolmatoff established this relationship between culture and environment on the basis of the Kogui's conceptual and philosophical principles.

> The Mother, and the indigenous idea of fertility, is at the center of what has been set out on the Kogui thinking; both concepts are supported by traditional principle and, even though this idea has been questioned from different perspectives (See Bocarejo 2001, Uribe 1990), it is a stronly held interpretation that defines the present Koguis as heirs of the Tairona archeological culture. Likewise, current concepts about Mother and Fertility are combined to form the *essence* that apparently defines the Kogui culture as an ancestral thinking mode whose spirituality guides a particular way of relating in a sustainable manner with the SNSM's natural setting. (Orrantia 2002:8)

In the 1970s, Gerardo Reichel-Dolmatoff (1985) asserted that Kogui thinking was an answer to their cultural and ecological adaptation to a historical process of obtaining food that found expression in their social organization (a food/sex model). This perspective constituted a model to be followed in future ethnological analysis. This view on the Kogui relationship with their environment also motivated the interest of environmental movements in the 1970s that led to the discovery of the Lost City.

Environmentalists have used anthropological descriptions to construct a "Kogui ecological image" within the transnational context (Orrantia 2002). This does not mean that Kogui people are not ecologically aware, but the point that I want to highlight is that independent of their actual ecological knowledge, anthropological discourses have helped to reproduce categorical schemes of nature and culture that have contributed to the Kogui's representation as *ecological natives* (see Ulloa 2001). This representation also resulted from the media boom created by the Lost City's discovery and the national impact of environmental movements, which promoted forest reservations and identified the SNSM's plants and animals as "relics" to be saved. The idea of preserving the SNSM's plants and animals was eventually extended to include the preservation of its "natives." From this union of environmental discourses and the media boom associated with the Lost City an *essentialized ecological native* arose that found further support in anthropological discourses addressing the continuity of a Tairona tradition and contemporary Kogui culture.

National and transnational conservation strategies have also contributed to the representation of the Kogui people as *ecological natives*. For example, the declaration of a National Park SNSM (1977) superimposed a parkland conservationist ethos on the Kogui *resguardo* that follows the policies of the National Administrative Unit of National Natural Parks. In addition, when UNESCO initiated its Man and the Biosphere program n 1979, it designated the SNSM as a biosphere reserve in accordance with UNESCO's mandate to protect unique land or coastline areas and promote environmental strategies that maintain biodiversity in the context of sustainable economic use.

The construction of the Kogui's ecological identity also occurred as a result of their own political activities as evident in the Organization Gonawindúa Tayrona's policies. From its formation, it has emphasized environmental issues, as expressed in the organization's constitution on January 1987. There it was agreed that "studies attempting to preserve ecology do not pose any problem" because the objective of indigenous peoples is to defend the SNSM. Later, the OGT gradually consolidated this relation when it established alliances with other national and international institutions that emphasized the organization's environmental interests.

In this context, the environmental NGO, FPSNSM emerged in 1986. Since its beginning, it has developed a series of activities that monitor local environmental conditions from stations located in the SNSM with the participation of peasants and indigenous peoples. FPNSM began its actions "in October, 1995 when it started a Rapid Ecological Evaluation (EER in Spanish) on the Sierra Nevada whose objective was to identify critical areas and to preserve ecological diversity, beginning with secondary information collection and analysis, remote-sensor image interpretation and updating social economic information" (www.prosierra.org). Then it consolidated these processes in the PDS that integrated the participation of governmental and non-governmental organizations and local agents, including the indigenous people and specifically the OGT (see chapter 3). This NGO has been important in publicizing the social and ecological situation of indigenous peoples in the SNSM, particularly the Kogui people. It has done so by employing a biodiversity paradigm to form its discourse concerning the SNSM and indigenous peoples. Accordingly, the FPSNSM promotes the view that the preservation of the SNSM has been thanks to its inhabitants, the indigenous peoples.

In these contexts indigenous peoples have begun to be valued due to their ecological knowledge, which has promoted the dissemination of indigenous peoples' *ecological native* image in global environmental circles, specifically the image associated with the Kogui people. All of these situations have also allowed the FPSNSM to increase worldwide interest in the SNSM and to raise funds both for its preservation and its "sustainable development."

There have also been agreements and financial support from multilateral institutions (World Bank, European Union) and national NGO's to implement the PDS. These were also largely the result of FPSNM's advertising campaign to legitimize indigenous peoples' ecological knowledge as an environmental conservation strategy.

In 1992 the OGT established an agreement with governmental institutions to develop the *Eco Sierra Plan* to promote *resguardo* enlargement and the "preservation and protection of the river basins that supplied water to over 10 municipalities and towns located in the flat parts of the Sierra's surrounding areas" (OGT 1998). This process gave a new dimension to indigenous peoples' territory because the ultimate objective was the region's protection and permanent water supply. The plan proposed to resolve deforestation problems due to drug trafficking and settlers' activities and set out joint actions between indigenous groups and governmental institutions in order to generate regional action.

In 1994 *Ricerca e Cooperazione* financed the Sierra Nevada Project. This project attempted to integrate rural development that was ecologically and culturally compatible and sustainable with the beliefs and activities of the indigenous peoples in the SNSM. The plan called for the integration of the economic and social concerns of the SNSM as necessary components for environmental preservation and management.

At the international level, Juan Mayr (Colombian minister of the environment 1998–2002) consolidated the linkage between indigenous peoples and environmental issues as a result of his international recognition through memberships in environmental institutions (UICN) and the Goldman Environmental Prize[4] (1993), which helped to position the SNSM's indigenous peoples as environmentalists. Juan Mayr founded the FPSNSM in 1986, serving as its executive director and helping to consolidate the preservation strategy for the SNSM until he was appointed minister. As the Goldman Environmental Prize committee explained:

> Mayr is a photographer who became an environmentalist, who lived two years in the Sierra Nevada with the Kogi, one of the last pre-Columbian civilizations that still exist. The Kogi, who live in considerable high-altitude settlements in the Sierra Nevada, adore knowledge and believe themselves to be the "Elder Brothers" of mankind. They hold themselves responsible for the universe's balance and look at the rest of mankind as their "Younger Brothers" who in their ignorance and greed are destroying their Great Mother (Earth). The Kogi, whom Mayr promotes as exemplary managers of the environment, have resolved to have minimum contact with the outer world (excerpt from the Goldman Environmental Prize).

The environmental problems of the SNSM have in fact been a matter of discussion since the 1970s, when specific projects were established through the OGT for their resolution. However, it was only in 1993 that environmental concern focused itself in a more systematic and regional manner in response to the deterioration of this unique ecosystem and its diverse plants and animals. Of particular concern was the lack of enforcement of national and international environmental policies. Deforestation, the introduction of inappropriate agricultural practices and inadequate technologies in addition to unlawful cultivations (coca and marijuana) contributed to this deterioration. These problems acquired the character of a crisis when their effects became evident in the loss of the water resources that affected not only the aqueduct supply in the main municipalities in Cesar, Guajira and Magdalena, but also the economy of these departments. Therefore, the FPSNSM generated a strategy to preserve the SNSM's biodiversity that addresses natural-resource preservation as a priority due to its impact on the region's economic interests.

In this way an "awareness" process started. The FPSNSM led projects to inform indigenous peoples and peasant communities of the necessity of recovering and preserving biodiversity in a manner aimed at sustainable development. These projects included training forums, workshops and symposiums promoting a balance between the environment and development. Preventing ecological destruction was promulgated as a means of defeating "underdevelopment" and "poverty" since the optimal use of resources and regional development would, in theory, increase along with conservation of the environment. The FPSNSM hoped that this program would improve the communities' "well-being" for the long term. Finally, another one of the FPSNSM's underlying concerns was to maintain the SNSM's biodiversity as an economic resource for the nation.

With this purpose, the national government set forth policies to stop the SNSM's environmental degradation, and so the government consolidated the PDS and Environmental Council (Consejo Ambiental Regional). These policies resulted from consultations, workshops, and campaigns that called upon participation from peasants, indigenous peoples, and economic and institutional sectors linked to the administration and management and sustainable use of natural resources of the SNSM. However, indigenous peoples have questioned this process and their conditions of participation (see chapter 3).

In 1998 the national government formalized the PDS within the national environmental policy guidelines consisting of two projects: LIL and GEF. The plans' global objective is to "conserve, protect and recover the SNSM's natural base to guarantee cultural, ecosystemic, and water source's preservation for the region's sustainable development" (100).

The various examples noted above demonstrate the consolidation of indigenous peoples of SNSM within the arena of transnational environmentalism. As Francesco Vicenti, member of the United Nations, said in the 'Forum for the Sustainable Development of the Sierra Nevada' (*El Informador,* May 20, 1998): "indigenous peoples' ancestral wisdom has given us harmony and conservation lessons."

However, the process of consolidation has not been totally consistent with indigenous values and interests. Despite the active and decisive participation of indigenous peoples in the PDS, they have often expressed dissatisfaction with its processes and goals. This framework greatly influences nonindigenous responses to indigenous peoples' claims, especailly development plans such as the PDS, which reflect nonindigenous economic conceptions of the environment that conflict with indigenous cultural relationships to their territories. The CTC proposes that environmental concerns of the SNSM have to be approached with the understanding that indigenous demands regarding cultural identity and political autonomy and self-determination are based on their territory. The indigenous peoples see these demands as the issues most in need of national and transnational attention.

INDIGENOUS PEOPLES' DEMANDS RELATED TO ENVIRONMENTAL ISSUES

Indigenous peoples around the world have often made declarations and demands regarding their cultural difference and distinctive identities. However, since Rio-92, their demands and declarations began to address environmental discourses and new conceptions of territory and political autonomy based on them. Diverse indigenous peoples around the world have decided to express through declarations of international scope their rights to manage their territories and natural resources according to their own traditions. Among these diverse declarations,[5] I have selected some that have had greater repercussions due to their emphasis on self-determination, territorial claims, resource management, alimentary security and traditional knowledge.

The *Kari-oca Declaration*[6] of 1992 was the first public and transnational pronouncement of indigenous peoples' conception of their relationship with the environment. This declaration was subsequently ratified at meetings of indigenous peoples such as those in Bali (June 2002) and Johannesburg (September 2002). This declaration is fundamental because it clearly states the philosophical bases of indigenous peoples' relationships to the environment. It gave voice, in a poetic way, to the thinking that sustains the connection of indigenous peoples to their environments: the relationship between culture and territory based on their ancestral traditions, which sustain their rights to self-determination and autonomous control of their territories and resources.

Kari-oca Declaration

> We, the Indigenous Peoples, walk to the future in the footprints of our ancestors.
> From the smallest to the largest living being, from the four directions, from the air, the land and the mountains. The creator has placed us, the Indigenous Peoples, upon our Mother the Earth.
> The footprints of our ancestors are permanently etched upon the lands of our peoples.
> We, the Indigenous Peoples, maintain our inherent rights to self-determination. We have always had the right to decide our own forms of government, to use our own laws, to raise and educate our children, to celebrate our own cultural identity without interference.

> We continue to maintain our rights as peoples despite centuries of deprivation, assimilation and genocide.
> We maintain our inalienable rights to our lands and territories, to all our resources—above and below—and to our waters. We assert our ongoing responsibility to pass these on to the future generations.
> We cannot be removed from our lands. We, the Indigenous peoples are connected by the circle of life to our lands and environments.
> We, the Indigenous Peoples, walk to the future in the footprints of our ancestors.
> Signed at Kari-oca, Brazil on the 30th Day of May 1992
> Reaffirmed at Bali, Indonesia, 4 June 2002.
> Reaffirmed at Johannesburg, September 2002.

In June of 1993 in New Zealand, indigenous peoples wrote *The Mataatua Declaration on Cultural and Intellectual Property Rights of Indigenous Peoples.*[7] This declaration asserted "the value of indigenous knowledge, biodiversity and biotechnology, customary environmental management, arts, music, language and other physical and spiritual cultural forms."

> **Declare** that Indigenous Peoples of the world have the right to self determination and in exercising that right must be recognized as the exclusive owners of their cultural and intellectual property; **Acknowledge** that Indigenous Peoples have a commonality of experiences relating to the exploitation of their cultural and intellectual property; **Affirm** that the knowledge of the Indigenous Peoples of the world is of benefit all humanity; **Recognize** that Indigenous Peoples are capable of managing their traditional knowledge themselves, but are willing to offer it to all humanity provided their fundamental rights to define and control this knowledge as protected by the international community; **Insist** that the first beneficiaries of indigenous knowledge (cultural and intellectual property rights) must be the direct indigenous descendants of such knowledge; **Declare** that all forms of discrimination and exploitation of indigenous peoples, indigenous knowledge and indigenous cultural and intellectual property rights must cease (The Mataatua Declaration on Cultural and Intellectual Property Rights of Indigenous Peoples, 1993).

Such indigenous proposals about cultural and intellectual property rights establish their right to define and decide their own protection mechanisms insofar as the existent ones are insufficient. They also propose a code of ethics that nonindigenous researchers can observe "when recording (visual, audio, written) their traditional and customary knowledge." Likewise, they promote the development of their traditional knowledges related to environmental and cultural practices, and finally, the recognition and recovery of their ancestral territories. They also prioritize the recognition of indigenous peoples as guardians and creators of their own knowledge and cultural traditions and their right to protect and to control their dissemination. The declaration also expresses the need to promote collective and individual property rights in a retroactive and historical way as a means of enabling indigenous peoples to fulfill their obligation to serve future generations by protecting them against cultural devastation.

Such proposals promote indigenous philosophical ideas of the relationship between indigenous peoples and the environment based on integral relations among territory, culture, knowledge and natural resources. Therefore, they cannot separate flora, fauna and people as independent things or objective commodities, and they demand a "moratorium on any further commercialization of indigenous medicinal plants and human genetic materials". For this reason, they insist on seeking appropriate mechanisms for the protection of their knowledge from within their cultural values and traditions. Finally, they suggest, as an indispensable condition, that the members of the United Nations ensure the participation of indigenous peoples and recognition of indigenous declarations, and that they generate sanction mechanisms for the states that violate the cultural integrity and property rights of indigenous peoples.

In 2001, at Isla Margarita, Venezuela, indigenous peoples participated in the IV Regional Workshop of the Andean Community on access to genetic resources, traditional knowledge and distribution of benefits. Nineteen representatives of indigenous peoples[8] established in their meetings the *Manifest of Indigenous Organizations Participants* that asserts that they are owners of the traditional knowledges that enable them to sustain their life and their territory as a whole and that that position is not negotiable. They also demand self-determination with respect to the right to decide the use of their knowledge. They state that the best way to guarantee self-determination is the recognition of their property rights over their territories and lands and their natural resources and biodiversity. Therefore, they demand a moratorium on development until clear strategies of protection of property rights and traditional knowledges have been established.

In 2002, in Guatemala, indigenous peoples[9] wrote the *Declaration of Atitlán. Indigenous Peoples' Consultation on the Right to Food: A Global Consultation.* This declaration, although it is more specific in regard to alimentary security, also states the close and key relationship between indigenous peoples with their territories and resources. The right of indigenous peoples to food is related to their sovereignty over their natural resources and their territories and to the autonomous decision to have their own policies on production, distribution and consumption according to their cultural and environmental practices. In this declaration, indigenous peoples establish that the right to food:

> [is] a collective right based on our special spiritual relationship with Mother Earth, our lands and territories, environment, and natural resources that provide our traditional nutrition; underscoring that the means of subsistence of Indigenous Peoples nourishes our cultures, languages, social life, worldview, and especially our relationship with Mother Earth; emphasizing that the denial of the Right to Food for Indigenous Peoples not only denies us our physical survival, but also denies us our social organization, our cultures, traditions, languages, spirituality, sovereignty, and total identity; it is a denial of our collective indigenous existence. (Atitlán Declaration 2002)

However, they also state that alimentary security has been affected by external processes such as: globalization and the neoliberal free market, displacement from their territories and the industrial exploitation of their natural resources, the imposition of industrialized models of development (for example, monocultivation and introduction of pesticides and chemical fertilizers), the lack of recognition of the intellectual property rights that indigenous peoples have in their natural resources, armed conflicts, and international policies aimed at controlling illicit cultivations. Therefore, they demand the adoption of the original text of the Draft Declaration on the Rights of Indigenous Peoples-UN, the ratification of the Convention on the Elimination of Persistent Organic Pollutants and the Kyoto Protocol on Climate Change and the ILO-169.

Indigenous peoples made interesting changes in this declaration because they assumed commitments for the achievement of the previous points. Some of these commitments are: to revitalize the worldviews of indigenous peoples; to strengthen their traditional food production systems and family and community economies; to create networks among the indigenous peoples; to pursue constructive ties with civil society; to promote autonomous indigenous processes directed toward the development of innovative systems for the protection of their knowledges, needs and worldviews.

The indigenous peoples synthesized their proposals in the *Kimberley Declaration,* a result of the International Indigenous Peoples' Summit on Sustainable Development held in Khoi-San Territory, Kimberley, South Africa, August 20–23, 2002. In this declaration the principles of the *Karioca Declaration* are reaffirmed, especially the indigenous peoples' relationship with Mother Earth and their responsibility to coming generations. Likewise, they reaffirm their right to self-determination and the political autonomy to control and manage their territories, lands, waters and resources. They demand the right to "determine and establish priorities and strategies for [our] self-development and for the use of our lands, territories and other resources. We demand that free, prior and informed consent must be the principle of approving or rejecting any project or activity affecting our lands, territories and other resources." Finally, they demand the recognition of the contribution of their knowledge to environmental and human sustainability. Therefore, their knowledge should be respected, promoted and protected, and their intellectual property rights should be guaranteed.

> The national, regional and international acceptance and recognition of Indigenous Peoples is central to the achievement of human and environmental sustainability. Our traditional knowledge systems must be respected, promoted and protected; our collective intellectual property rights must be guaranteed and ensured. Our traditional knowledge is not in the public domain; it is collective, cultural and intellectual property protected under our customary law. Unauthorized use and misappropriation of traditional knowledge is theft. (Kimberley Declaration, 2002)

In Colombia, indigenous peoples are in dialogue with TEIM and other indigenous peoples' movements. Indigenous leader such as Lorenzo Muelas,[10] Leonor Zalabata[11] and Antonio Jacanamijoy have already participated in various international meetings. In these meetings, apart from territorial recovery and demands for new political arenas and autonomy, there has been a clear emphasis in discussions on protection of the environment in which indigenous peoples' wisdom is recognized as a safeguard of the balance between humans and nature.

In the CPC-91, indigenous peoples also expressed their proposals in relationship to environmental discourses. The discussion topics were ecology and the defense of the environment based on indigenous peoples' knowledges and the proposals affirmed in the CPC-91 (art.330), which asserted: "the natural resources will be used without affecting cultural, social and economic integrity of indigenous peoples' communities."

In 2002, Lorenzo Muelas, as a candidate for the Colombian senate, presented a proposal to defend Mother Earth and all its waters, plants and animals, as the most important means to sustain human life. His slogan was "love your earth as your life." He stated, "I think that on this topic indigenous peoples are particularly qualified to contribute as inhabitants of this planet, because we know how to conserve, to manage and to use wisely by maintaining in balance all this wealth that nature is offering us" (Lorenzo Muelas' political brochure for the senate 2002–2006). These explicit articulations of politics with the environment, especially through the CPC-91 and indigenous candidates' proposals, have raised awareness of indigenous peoples' relation to the environment in international and national arenas.

In the Colombian context, there is no unified national indigenous peoples' declaration addressing environmental concerns, but rather diverse positions according to the point of view of specific indigenous leaders or organizations. These positions are indeed diverse, and they range from an outright moratorium on development to limited access to genetic resources under particular legal conditions (that is to say *sui generis* systems).

The lack of a unified position at the national level is due to a variety of factors: the need for more information in order to make decisions based on the principle of free and prior informed consent; the delay of the government in diffusing information and carrying out consultations with all the ethnic groups and local communities; individual and organizational perspectives that are not representative of all indigenous peoples and the limited opportunities for indigenous peoples' participation.

However, in the national context there are some individual positions that have gained support from other indigenous leaders and organization such as those of Lorenzo Muelas, UMIYAC and COICA. The last national meeting of the "Congress of Indigenous Peoples of Colombia" (2001) emphasized positions and proposals that highlighted the need to generate the information needed to initiate coordinated discussions in local, national and transnational contexts.

Lorenzo Muelas (1998) states that the relationship of indigenous peoples with their environment and its transformation have to be seen as a result of the indigenous peoples' knowledges and innovations that have generated their collective ways and understanding of identity. He states that "it is not possible to separate knowledge in which natural resources and peoples are already interrelated as some people think when they assume that nature is wild, or to divide the tangible from the intangible" (173). Likewise, he highlights how indigenous peoples see the universe as interconnected and unified so that all natural parts, even humans, are in permanent interrelation. Therefore, he states that to cut the world into parts and to privatize it is to violate the integrity of both natural and human life.

Lorenzo Muelas also discusses how the international law related to biodiversity and access to genetic resources has been imposed on indigenous peoples' territories to give business free access to their plants and animals. He concludes that this form of expropriation and exploitation is a continuation of the one that began 500 years ago. Only now it is the biodiversity in indigenous peoples' territories that is "coveted" by nonindigenous peoples, not gold.

Muelas calls for the organization and defense of indigenous peoples and their territories through the collective protection of their Law (*Derecho Mayor*) and their knowledges. Such protection implies that they will "not allow access [and demand] a stop to all research activity and gathering of resources and knowledges inside our territories" (1998:177).

In June of 1999, the Union of Yagé Healers of the Colombian Amazon (*Unión de Médicos Indígenas Yageceros de la Amazonia Colombiana-UMIYAC*)[12] made the "The Yurayaco Declaration." In this declaration, they highlighted the importance of yagé and medicinal plants in the construction of their knowledge, and they called for the recognition of their territories and collective intellectual property rights.

> We consider yagé, our medicinal plants and our wisdom, to be gifts from God and of great benefit for the health of humanity. This gathering may be our last opportunity to unite and defend our rights. Our motivation is not economic or political. We are seriously determined to demonstrate to the world the importance of our values. As sons of the same Creator and brothers and sisters on Mother Earth, we wish to speak, to offer our contribution so that life, peace and health may be possible. (The Yurayaco Declaration, 1999)

In this declaration, they establish the importance of using yagé for the Amazonian indigenous peoples' health. Likewise, they denounce nonindigenous peoples because they profane their culture and their territories by commercializing yagé and other plants. Therefore, they demand respect and recognition of their collective intellectual property rights, their territories, their medicine and their traditional healers or *taitas*. They also ask for the recognition of their medicine as a system of knowledge based on their cultural practices, traditions and history. Finally, they demand the autonomy to use their medicine and recover their sacred territories for the free exercise of their practices.

COICA[13] has generated a series of documents about the relationship of indigenous peoples and their environment in general and about access to genetic resources in particular. In the resolution of January 20, 2002, they state that, based on their rights of self-determination, only the indigenous can make decisions about the activities and projects that affect their territories. Therefore, they reject the Plan Colombia proposal for the development of indigenous peoples' territories for its environmental, social, cultural, economic effects and as a violation of human rights in its reduction of their sovereignty. They also call for governments and corporations to consider the adverse environmental and human effects of petroleum exploration and mining in their territories. They consider Plan Colombia a threat to indigenous peoples because its conception of development could cause the extermination of traditional cultures.

In relation to the environment, the indigenous peoples believe they have a fundamental role in defining how to balance the integrity of the environment with the demands for development in the countries where they live; and they believe that they should be the main agents of the recognition, defense and application of their fundamental rights and the decision-making processes that affect them and their territories. Therefore, they adopted the indigenous peoples' Amazon Agenda in its first phase, 2002–2005. The goals of this agenda are: sustainable development, indigenous peoples' rights, natural-resource management, education, political invigoration, and legalization and protection of their territories.

With regard to access to genetic resources and traditional knowledges, in 1999 the COICA proposed among others things: to incorporate the concept of collective cultural heritage of indigenous peoples into environmental policies; to establish special legal systems for the protection of indigenous peoples' knowledges; to value the innovations and traditional practices of indigenous peoples as informal innovations; and to recognize indigenous peoples as such and to avoid individual agreements regarding access to genetic resources.

Finally, in November of 2001, in the "Congress of Indigenous Peoples of Colombia," participants and representatives of several indigenous peoples generated three declarations titled "Planning of Territory and Autonomy are the Bases for Peace in Colombia," "A New Country, Worthy, Fair and at Peace; A New Government for Peace," and "Facing the War: Indigenous Peoples' Resistance and Peace for Colombians". In these documents they established the positions of indigenous peoples with regard to their territorial autonomy and political self-determination, and to their rights to define what to do with their resources or produce in their territories; all of which resulted in proposals for the construction of a new nation based on respect and democracy (see the webpage: www.onic.org.co).

Indigenous peoples of the SNSM are also part of these processes, and in their demands they also propose respect for their territory as well as political autonomy and self-determination. The CTC proposes to develop their notion of territory based on The Original Law and their own proposals related to environmental and territorial knowledge (as we have already seen in chapter 3).

In general, in the proposals of the representatives and organizations of indigenous peoples, at both the international and national levels, there has been a demand for the recognition of the links between their political autonomy and self-determination and their territories and natural resources. These proposals in the Colombian national context confirm what has been expressed in the international context related to the conceptions that sustain the relationship between indigenous peoples' identities and the environment. They agree that the bonds between territory, culture, knowledge and natural resources are basic to understanding indigenous peoples' relationship with their environment and their distinctive identities. Likewise, they demand the coordination of national and international policies with indigenous peoples' interests in a manner that respects the CPC-1991 and other international agreements.

CONCEPTUAL BASIS OF THE INDIGENOUS PEOPLES' DEMANDS

During the last two decades, indigenous peoples' have responded to governmental and nongovernmental institutions' environmental policies in national and international arenas. Indigenous peoples articulate their environmental proposals on four principal conceptual axes: the relationship of culture and territory, autonomy and self-determination, communal life plans for the future, and the right to food (alimentary security). These axes are interrelated and complementary, but I present them here separately.

The philosophical foundation that sustains these four axes is the notion of nature that establishes bonds among territory, culture, identity, knowledges and natural resources as an integral unity. Indigenous peoples' knowledges envision nature and societal relations as reciprocal. For indigenous cultures, nonhuman beings have social behaviors, and they are regulated by social rules. This assumes the idea of nature as a unity of humans and nonhumans which finds expression as a constant process of transformation and reciprocity. And although indigenous knowledges and their nature/culture relationships do not correspond to culture western categories, they have been in a dynamic relationship of interdependence with them for centuries (See for example: Descola and Pálsson 1996, Ulloa 2001a, 2002a, Milton 1996, Gragson and Blount 1999, Nazarea 1999, Grim 2001, Ellen and Fukui 1996, among others).

In general, indigenous peoples believe that nature is a live entity with agency, which can give permission to use its fauna or flora. Mother Nature (Mother Earth) is a being with which every one can talk and maintain relationships of reciprocity. They rethink the modern western notions of nature because they do not conceive of nature as an external entity, and so they do not have the modern western dualism of nature and culture. Therefore, they declare that decisions about biological resources cannot dispense with their notion of the unity of nature and people.

The conception of nature that informs the relationship of indigenous peoples and their environment fits well within one of the general tendencies of the global environmental movement called alternative holism or the biocentric paradigm (as we have seen in chapter 4). In fact, indigenous peoples' holistic conceptions of nature have acquired importance in diverse philosophic discourses: anti-industrial romanticism, antimodernism, spiritualism, social ecology, populism and ecofeminism, among others.

In particular, indigenous peoples around the world are positioning their environmental conceptions and practices as alternatives to the current western developmental and industrial approaches to environmental problems. They assert that nonwestern systems of knowledge, specifically indigenous knowledges, present alternatives to the intellectual assumptions that have led to the current environmental crisis.

This conception of nature sustains indigenous peoples demands for territory, self-determination and autonomy (autonomy is also related to continuity with pre-Columbian peoples, especially their connection to Mother Earth) and the right to control, plan and receive the benefits of their cultural knowledges and natural resources (life plans and alimentary security). In Colombia, these conceptual axes are present in indigenous peoples' demands in the following way:

Territory

Indigenous peoples' environmental objectives in their political actions are guided mainly by the defense of their territories. They always refer to their territories with terms such as "Mother Earth," "ancestral territories" or "sacred sites." Indigenous peoples' discourses emphasize territory as the central axis upon which they reproduce their lives. Recovery of their territory is a central struggle because as they recover more lands, they recover their cultural autonomy and ability to manage their natural resources. Therefore, it is of vital importance for all indigenous peoples to reclaim lost territories and make them traditional again.

Territory is constituted as a primordial factor of indigenous ethnic and cultural identity:

> Indigenous peoples' identity is based on three elements: the earth, language and thought. We all have our origin in the Mother Earth and the confluence of all elements constitutes our autonomy. However, we have lost our autonomy because of the exploitation and expropriation that the nonindigenous made upon our Mother Earth. They make us forget or be embarrassed by our language and they impose on us a foreign thought. (Encuentro de pueblos indígenas de la Costa Atlántica-1996)

They also state that to maintain the relationship between indigenous peoples and the environment it is necessary to have traditionally recognized territory.

> For indigenous peoples' worldview these concepts [territory, language and thought] are real. Therefore, knowing the situation of our territories it becomes urgent to enlarge the existent *resguardos,* as well as to constitute new ones. Indigenous peoples also have the responsibility of being guardians of natural resources, of the environment and of protected areas for the nation. (Encuentro de pueblos indígenas de la Costa Atlántica-1996)

Gabriel Muyuy who was a senator in 1991 and is now the ombudsman for ethnic affairs, points out that the recognition of indigenous peoples' political rights and autonomy is urgent, but to gain such recognition effectively, it is also necessary to work on issues regarding territory and the general protection of the atmosphere and natural resources.

In a Colombian national workshop, Indigenous Peoples, Traditional Communities and Environment–2002,[14] indigenous peoples and Afro-Colombian participants noted that discussion of their cultural, social and economic situations often requires a focus on the link between territorial and historical contexts. These contexts give a clear view of cultural and ancestral memory as the processes that have constructed the linkage between territory and identity among these indigenous and Afro-Colombian peoples. Therefore, they believe that making decisions on issues related to territory must involve the historical context in which they emerge. Consequently, when establishing such a cultural context and ancestral memory, they keep two concepts in mind, territory and self-determination, and it is upon their interdependence that indigenous people found their identities.

The notion of territory refers to a symbolic-cultural universe that transcends the level of physical things. These elements are of vital importance because the national government sometimes makes legal decisions that separate the legal territory of indigenous peoples from their territories of traditional and symbolic use. In the case of the indigenous and Afro-Colombian peoples, territory is not only a physical space but also a space of cultural meanings and actions. Therefore, they consider of vital importance the definition of territory developed by the ILO-169 or Law 21 of 1991, which conceptualizes territory in a cultural sense. On this basis, they also assert that self-determination implies the right of the ethnic groups and local communities to determine the use of their territories and resources in order to create the future that they want.

Autonomy and Self-determination

Indigenous peoples' dialogue with the state demonstrates their interest in integration rather than separation from Colombia and the world. This articulation of the indigenous peoples' territorial plans with those of the nation-state indicates, on the one hand, the indigenous desire for autonomous political authority and environmental control and, on the other, the desire for cooperation with the state in issues related to defining and managing territories and resources.

Indigenous peoples of Colombia, through their document titled "A New Country, Worthy, Fair and at Peace; A New Government for Peace", state that they want a presence inside the state, but in an autonomous way. They propose "to impel the self-government of peoples and communities, in the framework of the political sovereignty of the nation-state; which implies a guarantee of the autonomy of the indigenous, Afro-Colombians, Creoles and Rom peoples, and the autonomy of their social organizations. We speak of a society of radical democracy, which means self-determination of the peoples in their territories" (see www.onic-org.co).

Indigenous peoples demand the recognition of various legal rights as the foundations of their autonomy and identity. For example, they demand free, previous and informed consent through Previous Consultation (Consulta Previa), which was lacking when the Embera-katio territories were expropriated for the construction of a hydroelectric dam project. They maintain that their territorial and cultural rights have been harmed in such western-oriented "negotiations." In addition, the Uwa protests exemplify the complaint against the government's projects that "harm Mother Earth, affecting the autonomy and the authority of our territories". Both peoples ask to be respected and that the government agencies or private companies that want to develop projects in their territories seek prior consultation agreements.

Indigenous peoples also assert that the state has discriminated against them and abandoned them by ignoring its promises and agreements. They want the government to commit financial support to their communities in order to help them defend their rights and promote respect for their customs and organizations. However, indigenous peoples' discourses may also be contradictory because, on one hand, indigenous peoples promulgate the defense of their territories from development projects, and on the other, they also protest because the government does not plan developmental projects to improve their conditions of life: highways, endowment of basic services, schools, and health centers.

Furthermore, indigenous peoples' territories are in the middle of the conflict between paramilitary and guerrilla forces that dispute indigenous control over them. Therefore, indigenous peoples also call for autonomy and self-determination because they do not want their territories to be battlefields, and they also demand recognition of their neutrality in such armed conflicts. Indigenous peoples have protested this "invasion" of their territories, and they forbid entrance of these groups into their territories. Anatolio Quirá, indigenous former senator and consultant of the CRIC, affirmed that "the force of the Indian is the force of the earth and toward her we won't allow more disregard."

Life Plans

Indigenous peoples express their demands for self-determination in their "Life Plans" which bind the cultural activities of indigenous peoples to the environmental processes of their territories. By doing so, indigenous peoples have become symbols of the defense of biodiversity and endangered ecosystems in the West's imaginary. Several indigenous peoples are planning diverse proposals for the ecosystemic management of their territories according to life plans that link future development to their needs and cultural practices, and so the life plans of indigenous peoples have become a legal strategy to defend their cultures by articulating their concept of territory to the West's concepts of environmentalism and biodiversity. Indigenous peoples state that such life plans are essential to cultural conceptions that guarantee the continuity of their identities through time.

For example, the OREWA's life plan titled: "What We Want and We Plan to Make in Our Territory" (Lo que queremos y pensamos hacer en nuestro territorio) founds itself upon the relationship between culture and territory.

> For indigenous peoples of the department [Choco] well-being is based on the sense of ownership of Nature like an ancestral legacy of origin. Multiple spirits constitute nature and the world and each one of them represents one of the species of the alive or dead beings that exist. The symbolic interrelations of the worldview of indigenous peoples play a definitive role in the balance of the ecosystems that are in our territory. It is for that reason that when there is harm to Nature there is harm to culture and vice versa (1996:423).

In the case of the indigenous peoples' grassroots movement called Organization of Antioquia (Organización Indígena de Antioquia-OIA), participants presume that the life plans have to be specific to each indigenous culture.

> We think that life plans are part of our self-determination. They are extremely important because they are related to social control, cultural crisis, alimentary security, and conditions of life that have to do with economic production, health and cultural polices related to bilingual education. Education should strengthen our relation with Mother Nature. (Workshop: Indigenous peoples, Traditional Communities and Environment. (ICANH-Institute von Humboldt-Colciencias, 2002)

The environmental politics of indigenous peoples' movements propose as vital for the conservation of biodiversity the legal recognition of their cultural definition of territories (according to the ILO-169 and the CPC-91) and the guarantee that the *resguardos* fulfill their ecological function. This assumes the articulation of their territorial proposals with national policies for sustainable development.

The Right to Food (Alimentary Security)

Alimentary security is another basic axis in the demand of indigenous peoples due to nonindigenous territorial demands and the imposition of development models that have disrupted indigenous knowledge and practices related to alimentary practices. Therefore, indigenous peoples have developed a series of strategies to recover seeds and the knowledges associated with them as strategies of resistance and cultural recovery. In the document generated in the congress of indigenous peoples titled “A New Country, Worthy, Fair and at Peace: A New Government for the Peace,” indigenous peoples demand economic, ecological and alimentary sovereignty that guarantees equal human relationships and solidarity: “ We speak of an equal society and of social justice, understanding that as the real and full exercise of social, cultural, economic and environmental rights and services” (www.onic.org.co).

Likewise, they demand the ratification of proposals presented in international arenas related to indigenous peoples: for example, Agenda 21, the ILO-169, and the Permanent Forum on Indigenous Issues. Finally, they focus their attention on transnational institutions and civil society as the arenas where they hope to promote their alternatives to western development as means to protect their alimentary security.

The implications of the declarations of indigenous peoples can be summarized in this manner: the opening of more democratic arenas for participation in which the indigenous peoples may claim to be environmental authorities, thus allowing them control over their territories and natural resources. Accordingly, they demand the implementation of policies that sustain the articulation of their territorial notions with their identities and that support their responsibility to promote environmental and cultural conservation. In this manner, they also promote a new dimension of global sustainable development by proposing local strategies that allow counterhegemonic globalizations. By challenging western conceptions of culture, territory and environment, the indigenous peoples have helped create new meanings and cultural practices that build, according to Leff (2002), “cultural territories” that promote alternative management strategies and forms of environmental conservation and cultural continuity.

FINAL REMARKS: INDIGENOUS PEOPLES' AUTONOMY, NEW NATION-STATES?

> *Article 3*. Indigenous peoples have the right of self-determination. By virtue of that right they freely determine their political status and freely pursue their economic, social and cultural development.
>
> (UN/Draft Declaration on the Rights of Indigenous Peoples, 1994).

The new interrelations between indigenous peoples and environmentalists have effectively situated indigenous demands in international political arenas. These demands are constructed around concepts of territory, self-determination and sovereignty that resemble the western nation-state model. However, indigenous peoples' struggles for self-determination imply the recognition of collective rights and a particular legal system within their territories that are distinctly nonwestern.

There are more than 300 million indigenous peoples around the world, and more than 40 million in Latin America who have been recognized by international law through, for example, the ILO-169), the Draft Declaration on the Rights of Indigenous Peoples (1992), and the CBD (articles 8j and 10c), among others. These national and international recognitions link indigenous cultural diversity and identity to transnational indigenous rights. In fact, international indigenous peoples' rights have forced the rethinking of western assumptions regarding sovereignty, national territory and local indigenous rights.

The American Indian Law Alliance's (AILA) statement on the right of self-determination asserts (December 3/2002):

> 1. The American Indian Law Alliance affirms its position that Article 3 of the draft U.N. Declaration on the Rights of Indigenous Peoples must not be altered, directly or indirectly, through amendments to or restructuring of the Declaration.
> 2. Even where limitations are not expressly referred to in the draft Declaration, the consideration of current international standards, as well as the competing rights of others, effectively impose limitations on human rights. "Any consideration of a human right must be in the context of current international standards. Taken together, the legal rules result in the position that there are limitations on most human rights in order to protect other rights and interests of society; these limitations are interpreted narrowly, but

consideration is given to the context of the specific society affected and of current international standards" R. McCorquodale, "Human Rights and Self-Determination" in M. Sellers, ed., *The New World Order: Sovereignty, Human Rights, and the Self-Determination of Peoples* (Oxford/Washington, D.C.: Berg, 1996) 9 at p. 15.

3. As virtually all participants in this Working Group acknowledge, the affirmation of the right of self-determination of Indigenous peoples is a core element of the draft Declaration and essential to its integrity. As provided in the international human rights Covenants, this right applies to "all peoples". Consequently, it would violate the peremptory norm prohibiting racial discrimination to create a different and lesser standard for the world's Indigenous peoples concerning this crucial human right."The major distinguishing feature of such rules [i.e. peremptory norms] is their relative indelibility. They are rules of customary law which cannot be set aside by treaty or acquiescence but only by the formation of a subsequent customary rule of contrary effect. The least controversial examples of the class are the prohibition of the use of force, the law of genocide, the principle of racial nondiscrimination, crimes against humanity, and the rules prohibiting trade in slaves and piracy." I. Brownlie, 5th ed., *Principles of Public International Law* (Oxford: Clarendon Press, 1998) at p. 515.
4. Clearly, the right of Indigenous peoples to self-determination is a "democratic entitlement." Denial of this right or the creation of a discriminatory lesser standard would be incompatible with true democracy. "Self-determination is the oldest aspect of the democratic entitlement . . . Self-determination postulates the right of a people in an established territory to determine its collective political destiny in a democratic fashion and is therefore at the core of the democratic entitlement" T. Franck, "The Emerging Right to Democratic Governance," (1992) 86 *Am. J. Int'l L.* 46, at p. 52.
5. Further, to undermine this right or other human rights of Indigenous peoples would be a violation of the Purposes and Principles of the Charter of the United Nations, which the U.N., its Member States and organs are all legally bound to respect. "The Purposes of the United Nations are: 3. To achieve international cooperation . . . in promoting and encouraging respect for human rights and for fundamental freedoms for all without distinction as to race, sex, language, or religion" Charter of the United Nations, Art. 1, para. 3

According to Griswold (1995), for some, indigenous peoples' claims are "interpreted as posing a threat to state sovereignty and territorial integrity" (94). Their claims produce a change from "the minimalist position of states, to the visionary one of indigenous peoples" (Lam quoted by Griswold 1995:95). Moreover, she describes how the recognition of these rights "reflect another version of the world than the one transmitted by European-based nationalism" (1995:96). Therefore, she notes how these rights, especially the right of self-determination, represent a challenge to static notions of the state when she describes how different "experts" believe that the recognition of indigenous self-determination could lead to secession or could sponsor new structures and processes within the nation that might qualify its sovereignty.

In a similar way, Kymlicka (1996) shows how, in Canada, a multinational and polyethnic state, the recognition of three different forms of differentiated citizenship (polyethnic rights, special representation rights and self-government rights[15]) generates questions related to social unity and national identity. He asserts that polyethnic rights and special representation rights are claims for inclusion rather than separation: "Groups that feel excluded want to be included in the larger society, and the recognition and accommodation of their 'difference' is intended to facilitate this" (162). He argues that the right of special representation may promote greater participation, equality and integration, especially polyethnic rights, which open a political space for ethnic minorities within public affairs in which they have not had effective representation. Nevertheless, he notes that self-government and autonomy "raise problems for the integrative function of citizenship" (163). Therefore, following Kymlicka (1996), the idea of a multicultural nation based on self-government rights also implies separation: "Self-government rights, therefore, are the most complete case of differentiated citizenship, since they divide the people into separate 'peoples,' each with its own historical rights, territories, and powers of self-government, and each therefore, with its own political community" (163).

Indigenous peoples' rights of self determination promote political practices based on communal participation and collective identity as alternatives to western models of individual citizenship and relations to the state. In addition, their demand for the recognition of cultural practices that depart from western notions of the public and private also promote different conceptions of the political sphere. Their ways of participation in and conceptions about public and private spheres have thus helped to open new discursive and practical opportunities for other social movements. Moreover, indigenous peoples' movements have opened political arenas not only to the possibility of rethinking the cultural differences between humans but also to the possibility of rethinking relationships between humans and nonhumans.

In this way, indigenous identities transcend the political idea of individual citizenship to incorporate broader social, cultural and natural considerations. Thus, the western legal notion of the unique citizen is transformed by the notion of "differentiated citizenship" (here I follow Kymlicka 1996) that can establish not only new relationships between indigenous peoples and governments within the nation-sate, but also new relations among all citizens and nature. As a result, indigenous peoples are creating a new way of doing politics that allows the rethinking of the meaning of citizenship, nation, and nature.

Indigenous peoples' self-determination rights also imply the recognition of cultural practices and territories that challenge notions of national and transnational borders. Thus, for the state, this could mean the fragmentation of its territory, while for the indigenous movements, however, this process implies the recognition of their sovereignty over their territories. In some cases indigenous peoples' territories transcend state borders because the indigenous notion of territory establishes relations with other indigenous cultures and international regimes (NGO's and ecological movements, among others), which places the relationship of the nation and indigenous peoples within transnational networks.

Indigenous peoples transcend not only territorial borders (as do migrants and diasporas), but also cultural borders. In this way, indigenous peoples establish new political relations, different memberships and identities within international regimes, transnational social movements and indigenous nations. Moreover, indigenous peoples' practices show that transcultural and transnational encounters are integral to their cultural dynamics and identities. These multiple, dynamic identities and loyalties (conservationist, NGO's, and indigenous nations, among others) situate indigenous peoples' movements at the forefront of new conception of citizenship within or beyond those of nation-states.

The indigenous movements' actions have contested the law through the means of the law by negotiating and resituating themselves within national constitutions (Lazarus-Black and Hirsch 1994). Therefore, indigenous peoples' movements, as collective identities, have created a political arena to contest and challenge identities formulated within the official system. Moreover, indigenous movements have "manipulated" the legal system, not only using it, but also redefining it through different legal strategies such as free and previous informed consent, environmental and cultural assessment, the ILO-169 and the CBD, among others.

Although indigenous cultures have asserted their cultural practices and identities for a long time, it is only now that their identities, proposals and rights are being recognized as integral to national and transnational political arenas. These transnational relations have been called global, transnational, postnational or cosmopolitan, and the new political dynamics and political agents they sponsor have to be analyzed in a different way that requires rethinking and reconfiguring western conceptions of political communities, collective identities, processes of democracy and participation. Among these analyses, some promote alternative visions of multiculturalism, diversity, national territory and transnational realities based on a different conception of democratic participation.

However, there remains a question regarding the motives of the West's recognition of indigenous identities and all the changes they imply (see chapter 2). The West's recognition of indigenous peoples' identities is ultimately tied to an interest in their territories and resources, and much of western policy has been dedicated to transforming indigenous peoples and their territories into western-style political-economic "entities" (micro nation-states) that can be detached from indigenous cultural processes. Therefore, indigenous peoples and their territories can become things in an instrumental world of contracts and self-interested economic agents in which relations to the environment are technical practices without cultural context that can be formulaically implemented around the world.

Maurer (2000) working in the Caribbean, points out that the modern western process of mapping helped to construct new relations with the land that gave rise to increased western influence. A similar process is occurring in the SNSM through discursive innovation and the practical changes that follow from it. By engaging in western discourses and practices, the indigenous run the risk of assimilating to the order of things that they set out to change. Indigenous peoples' demands for self-determination can be co-opted, adjusted and reintroduced to indigenous peoples in a manner that even more effectively subordinates them to the values and interests that inform western governmental techniques: development in conformity with the principles of economic and political individualism and their technical application.

The indigenous identity-construction process addressed here defines western ideas of territory, autonomy, property rights and environmentalism in new ways and even offers alternatives to them in some cases. Nonetheless, the representations of indigenous peoples disseminated in the western media often obscure this newness from western peoples. In the next chapter, I focus on representations of *ecological natives* in order to develop the implications such representations have for indigenous peoples' identities and territories as well as the identities and environments of nonindigenous peoples.

Chapter Six

Environmental Images and Representations: Implications for Indigenous Peoples

INTRODUCTION

In the 1970s, indigenous peoples gained conceptual recognition as environmental figures in various national and transnational environmental discourses, representations and governmental policies. However, it is in the last decade that *ecological natives* have gained practical recognition by assuming leading roles in national and transnational contexts as environmental and political agents (as we have seen in the previous chapters). Representations of *ecological natives* have become important to the political strategies of both indigenous and environmental movements; and it is the way that each movement uses representations of the *ecological native* that generates the understandings and misunderstandings that affect the possibility of practical cooperation between them and the effective realization of their goals within the context of global eco-governmentality.

Environmentalists invoke the figure of the *ecological native* as a powerful means to legitimize their discourses and practical aims, just as indigenous peoples use the figure to legitimize the cultural values and environmental practices and policies evident in their own discourses and activities. (Ulloa 2001). Indigenous and environmental movements share the *ecological native* as a means to focus their interests and goals in the legal and political arenas. And so, just like the legal and political issues surrounding the environment, the representations of the ecological native are always under negotiation, have inherent limitations in their scope, and produce contradictory relations to nature, culture, humans, property rights and natural resources (Flórez 2001). In short, the ecological native is currently among the most important shared modes of representation available to negotiate an understanding of the environmental crisis that faces humanity and to generate responses to it.

A brief example demonstrates how indigenous and environmental movements may employ the ecological native as a means to enable cooperation and goal sharing. In May 1996, in Costa Rica, the Earth Council and the Fetzer Institute held a meeting that produced the declaration *Indigenous Peoples, Mother Earth and Spirituality.*[1] The meeting established that relationships of indigenous peoples with their environment arise from natural laws that are necessary to maintain a balanced relationship with Mother Earth. The document calls for the recognition of indigenous territories, sacred sites and cultural and intellectual property rights as a means for all inhabitants of the planet to contribute to the continuity of nature for future generations.

> The knowledge of Mother Earth flows through our veins. We are children of the Creator and Mother Earth and we were instructed to live in balance and harmony with the sky, Mother Earth, plants, trees, rocks, minerals, winged ones, water beings, and the four legged. (Indigenous Peoples, Mother Earth and Spirituality Declaration, 1996)

What is noteworthy in the current context is that this document employs and synthesizes western and indigenous concerns about the environment in a highly metaphoric or poetic language that often characterizes indigenous statements, and it combines indigenous spiritually with western technology through the circumstances of the meeting and document's production and dissemination: air travel, web sites and modern bureaucratic organizations. It is an example of how the ecological native has established itself as a global presence and an important implement for mediating the interaction of western and indigenous values and interests. However, the ecological native is hardly a neutral or uncontested figure and, as I will demonstrate below, it has most often been used by western peoples as a means to dominate the indigenous peoples of Latin America rather than a means to mediate the differences of the two peoples.

Environmental discourses, whether of indigenous or nonindigenous origin, continually produce, modify and disseminate various stereotypes that affect national and international images and representations of indigenous peoples' identities, and many of these stereotypes have existed for centuries and have often exercised a negative influence over indigenous peoples. In the

West, there are many instances of discourses that employ notions of untamed nature and ideas of "primitivism" that stereotype indigenous peoples and nature as undeveloped and that put them in unequal power relationships to westerners. In others, they oscillate between representations of indigenous peoples as parts of nature or parts of culture, a specifically western paradigm that resembles colonial civilizing processes. In addition, some represent indigenous peoples as superior or as "noble savages" who provide westerners the values they need to complete the apotheosis of their own selves and societies.

Construction of identity is inextricably related to conceptions and representations of the self in relation to the Other. Representation is thus an essential concept for understanding identity. As Hall (1997) states, representation means "using language [verbal or nonverbal] to say something meaningful about, or to represent the world meaningfully, to other people" (15). This process implies two systems: first, a set of concepts or mental representations and their organization to interpret the world; second, a common language or set of signs (written words, speech or visual images) that allow individuals to communicate with a shared code. In order to produce meaning, these representations have to be located within historical processes and related to specific cultural contexts and situations. Consequently, representations of the Other necessarily involve issues of power, gender, nationality, class, race, age, status, ethnicity and sexual preference.[2]

Said (1997) makes an important point in his remark on the necessity of understanding the construction of the Other in a historical context. He proposes locating the individual, the social and cultural context, the representation, the interpreter, as well as the interpreted in their mutually determining historical roles.

> In other words, all knowledge of other cultures, societies, or religions come about through an admixture of indirect evidence with the individual scholar's personal situation, which includes time, place, personal gifts, historical situation, as well as the overall political circumstances. What makes such knowledge accurate or inaccurate, bad, better, or worse, has to do mainly with the needs of the society in which that knowledge is produced. (Said 1997:168)

Recent critiques of the colonial construction of the Other seem to satisfy the new politics of the recognition of difference. However, even the most self-conscious and self-critical version of the western imaginary still has stereotypes that mark and constrain the identity of the Other, and such is certainly the case of the *ecological native*. I argue that conceptions and representations of indigenous peoples as *ecological natives,* and therefore as parts of westernized environmental discourses, recall colonial strategies of representation of the Other (often employed as a means of territorial appropriation), and that they often ignore indigenous peoples as social agents by categorizing them as part of "nature." So too, such representations of *ecological natives* reproduce the dualist conception of nature and culture that causes contradictory visions of the indigenous and nature within western environmental discourses.

Although representations of the *ecological native* are multiple, global environmental discourses consistently promote two main ideas: one related to biocentrism and the other related to modernity that are complementary and that require one another in order to generate discussion, contradiction, opposition and agreement. These two ideas are the Other as an integral part of Mother Nature and the Other as a fellow human in need of expert western help to achieve sustainable development and self-improvement. Both representations reproduce colonial discourses of "helping" or civilizing by characterizing indigenous peoples as "savage natives," which also puts their territories under the inevitable idea of needing to be saved and protected (often from the indigenous inhabitants themselves).

To develop these points I present how, in Colombia, there have been three historical moments in the relationship of indigenous peoples and nature: the colonial (characterized by their naturalization), the republican (characterized by their culturalization) and the environmental (characterized by their re-naturalization). These three moments demonstrate a conceptual return to a colonial process of appropriation of natural resources, territories and humans in which indigenous peoples become "objects" of western knowledge, management and use. However, there are also spaces and ruptures in western representations and discourses that allow the indigenous to contest, appropriate and resignify their otherness.

Although I believe that indigenous peoples' ecological knowledges do have much to teach us about nature, I do not focus on the possibilities of these knowledges here. Rather, my aim is to critically examine the representations related to *ecological natives* within western environmental discourses. In this way, as Said suggests, the indigenous can help us learn what is good or bad in the reasons and ways we produce and reproduce our "knowledge" of them — the Other.

WESTERN EYES AND THE OTHER

The western construction of the Other has generated a way of classifying difference through specific relations of power in which the Other becomes an object of knowledge for the western epistemic. By doing so, western thought uses the notion of difference as a mechanism of power to mark, assign and classify otherness as an object of colonial understanding, control and assimilation. Western thought promoted colonization by elaborating different narratives of progress and order in scientific, political, economic, aesthetic and social discourses, among others, in order to explain such cultural differences. These narratives assumed a particular way of representing the Other that took form in practical activities, personal attitudes, texts, policies and objects that share the same assumptions and belong to the same "discursive formation." Missionaries, agents of empires as well as anthropologists played major roles in constructing this Other.

Western discourse frequently designates the Other's differences as indications of weakness, inferiority or backwardness. Very often, these differences have been explained through scientific knowledge (evolutionism, environmental determinism and diffusionism, among others) that locates the Other in a position of subordination. Lutz and Collins (1993), following Stocking, present how the differences manifested in social inequalities are represented as a result of "uneven biological or cultural development." As Said notes, such ostensibly empirical differences justify processes of colonization and domination which represent the cultures of the Other as static, self-contained and self-reinforced systems (Said 1978). Also, regarding racial differences, they show how "those in power elaborate observable physical differences—no matter how subtle—into explanations, affirmations, and justifications for inequality and oppression" (Lutz and Collins 1993:155).

Lutz and Collins' work (1993) places particular emphasis on how "National Geographic" magazine has represented the Other as exotic, sexualized and naturalized. In addition, they show how the magazine represents the Other as a timeless and passive inhabitant of his or her own situation. The Other occupies a fixed position in time and space, unlike the progressive and mobile self of the West which can transcend its own position to occupy that of "objective" truth or scientific neutrality. Discourses of the Other are based not only on such "objective" signs and images of otherness, but also on their absence. Consequently, the Other is often represented as being without history or any relation to western ideals of development, progress or industry (Nochlin 1989, Said 1978).

In addition, western representations often place all the different Others together in a common category, "the nonwestern world", thereby erasing the diversity that exists among them. As Lutz and Collins (1993) show in their analysis of National Geographic, western representations depict only two worlds: "the West and its technological and social progress" and the non-West of the Other. In a similar way, Said (1978) has observed how western texts represent the Orient as the opposite of the West, which is "rational, peaceful, liberal, logical, capable of holding real values, without natural suspicion; [and] the latter [Arab-Orientals] are none of these things" (49). In addition, Steiner (1995) shows how the discovery of the New World also promoted the European tendency to distinguish itself as normal relative to the strangeness of the Other.

> Whenever they were encountered, they were portrayed and represented by the same people—European observers who reduced them to a metaphor of Otherness that served only to confirm European expectations of the exotic rather than to challenge those assumptions (Steiner 1995:203).

Social scientists and their discourses have also played key roles in the construction of the Other. Such "scientists" helped the colonial processes by producing the "information" and perspective of authority those processes required. They produced a scientific or "expert knowledge" which could be useful not only in academic arenas but also in practical and political tasks. Most importantly, this "expert knowledge" conferred on Europeans the authority to define and locate the Other. Said (1978) has noted how such textual production generates a systematic knowledge about the Other whose apparent participation in the European metanarrative of scientific objectivity and neutrality vindicates the subordination of the Other and the exercise of European power through the colonial process.

Sciences such as ethnography, anatomy, philology and history quickly assumed the task of explaining the Other's differences (shortcomings). Orientalism thus explained the Orientals' behavior and gave them a mentality, a history and context. As Said notes, this discourse located the Other in an arbitrary geographical space and defined cultural and historical differences according to unique idea of representation—the closed and unchanging cycle of Oriental life. Orientalism not only erased the cultural particularities and agency of the peoples it represented, it also located Europeans in positions of power: those who had changed, advanced and evolved.

The stereotype of the identical nonwestern Other and its nonprogressive, unevolved or undeveloped life derives from specific practices of representation that create specific visual and textual stereotypes, which in turn create the homogenizing time and space of the Other in the western imaginary: for example, the noble savage or the primitive. In the case of indigenous peoples, becoming the object of expert knowledge has often contributed to their representation as primitives:

> To study the primitive is thus to enter an exotic world which is also a familiar world. That world is structured by sets of images that have slipped from their original metaphoric status to control perceptions of primitives—images and ideas that I call 'tropes'. Primitives are like children, the tropes say. Primitives are our untamed selves, our id forces—libidinous, irrational, violent, dangerous. Primitives are mystics, in tune with nature, part of its harmonies. Primitives are free. Primitives exist at the 'lowest cultural levels'; we occupy the 'highest,' in the metaphors of stratification and hierarchy commonly used by Malinowski and others like him. The ensemble of these tropes—however miscellaneous and contradictory—forms the basic grammar and vocabulary of what I call primitivist discourse, a discourse fundamental to the Western sense of self and Other. (Torgovnick 1990:8)

Experts have filled museums and collections with drawings, engravings, videos, texts, pictures and other materials that have fixed the Other in a new western context. Not only the objects, but the means and motives by which they were "appropriated and transferred, [. . .] provided the material out of which new stories about the world could be created" (Lutz and Collins 1993:149). As for the western spectators, these representations of the Other provoked comparison between different cultures that implicitly gave a superior, detached and, ultimately, transcendent position to them simply by subjecting the Other's culture to representation and making it an object of consumption. In this way, such representations of cultural superiority and hierarchy assumed an ostensible objectivity that allowed the spectator to remain innocent regarding the power relations that placed him or her in such a position of dominance: with the power to consume the Other.

The technical or mechanical reproduction of representations also served as a guarantor of the truth and objectivity of western representations of the other. For example, visual representations confirmed the ideal of the Other because these images were considered an objective testimony and proof of the Others' world. Visual narratives constructed the perception of the Other and justified processes of colonization insofar as such representations consistently reproduced the authority of the witness/observer and thus, through mechanical reiteration, they constituted proof of the reality of his/her superior-power and knowledge. Consequently, since the sixteenth century, the accounts of travelers used illustrations because "they added a very powerful element of realism to the narrative description of distant peoples and foreign lands" (Steiner 1995). That realism, however, was more a function of creating a consistent stereotype and sense of the "real" than it was of creating any sense of the difference or otherness that often marks contact with the real.

Moreover, many expeditions included an artist to register cultural differences and provide "a source of accurate and timely information on the colonial world" (Lutz and Collins 1993). With photography, the production of expert evidence achieved a level of technical independence from the observer and the observed that suggested that the representations produced had become the truly neutral "presentations" of reality. Images of Others were thus easily relocated and consumed, Lutz and Collins note, as "a direct transcription of a reality that was timeless, classless, and outside the boundaries of language and culture" (1993:28). (Ironically, these descriptions might better describe the conception of western superiority than the "reality" of the Other.)

Western visual images are cultural and historical constructions determined by a specific discursive formation. These representations reflect the intentions of the author who operates under specific discourses that motivate ways of seeing and representation that often exercise a hegemonic power that makes them invisible to the author. Nonetheless, visual perception is always mediated by cultural practices. This can be seen in how the colonial process and representations of otherness employed western categories of gender and race.

Steiner (1995) shows how during the 16th and 17th centuries illustrators repeatedly used in their engravings previous visual representations that became stereotypes of the Other that were continually reproduced over the course of centuries. When they could not go to the place, they borrowed elements from the stereotypes of various cultures to create a "conventional canon to represent the non-European subjects they had never seen" (Steiner 1995:209). In addition, the readers could not verify these representations except by comparing them for consistency with similar representations. Consequently, they came to expect a specific stereotypical representation consistent with their experience of the western cultural imaginary. Steiner notes how scenes of drumming and dancing (a barbarous spectacle) were a common representation of the "primitive" society during the 17th and 18th centuries. Similar representations were used to depict people from radically different epochs and cultures to suggest the sameness of their inferiority.

Such visual representations reproduced specific iconographic genres related to European ideals of the "primitive" that traveled from place to place synchronizing the imagination of European consumers and confirming their stereotypes and sense of superiority. In a similar way, Webb (1995) points out how the manipulation of photographs has been a common practice throughout the history of colonization. She describes how "the usc of retouching techniques and controlled settings by and for nineteenth-century Europeans was generally well suited to their canons of portraiture and representation" (176). Representations of the Other, in this case the Pacific Island peoples, were motivated by a desire to capture and affirm stereotypes of "primitive life." Photograpers retouched images to include body painting, weapons or clothes that designated peoples of the Pacific islands as exotic Others under Western categories of "primitivism."

Ideals of the Other have also been expressed in terms of gender relations, particularly in colonial discourses which used western patriarchal gender conceptions to represent the Other according to dichotomous conceptions of men/women, culture/nature and public/private, just to name a few. Representations of the Other in terms of gender relations assumed that patriarchal relations were normal in all societies. These representations assumed that nonwestern societies had western relations with nature in which women were thought to be closer to nature and men to culture. Such representations helped to reproduce the inequalities and stereotypes that still afflict the relationships of western to nonwestern people by equating the latter with western valuations of the feminine.

Current techniques for the production and consumption of representations of the exotic or primitive have reinforced and maintained western relations of power over Others. Moreover, among western people, these representations have become increasingly prevalent in daily practices. Conceptions of Others as "exotics" have accordingly become more effectively embodied and reproduced through *habitus* (Bourdieu 1993) or daily practices, such as the consumption of movies, videos, web sites, books, museum exhibitions, and art.

HISTORICAL PROCESSES OF INDIGENOUS PEOPLES' IMAGES AND REPRESENTATIONS

Talking about indigenous peoples' representations in Latin American, in general, and in Colombia, in particular, entails going back to the Conquest of America in order to give context to the European presence. This encounter produced changes not only for America but also for Europe insofar as the New World played an important part in the production of modern thought. Moreover, the process of the conquest of the New World produced not only a structural break in indigenous peoples' cultures, but also a new idea of the "Other" that forms one of the bases of the Renaissance humanist tradition.

During the 18th century, although the *Criollos* (Spanish-Americans) had a better social and economic position than indigenous peoples, *mestizos, mulatos, zambos*[8] and *black* peoples, the Spanish state also discriminated against hem and dominated them. The Peninsular Spanish held government positions, while the *Criollos* were excluded from positions of political power, though they were the most important local economic group. Therefore, by the beginning of the 19th century, Criollos in most of the Latin American colonies were struggling to found independent republics in order to put an end to the colonial order.

The Colonial Naturalization

In Renaissance humanism, the Other was considered part of humanity, in contrast to later conceptions that naturalized the Other in a rigid dichotomous hierarchy and universal order in which the European occupied the superior, civilized position. Therefore, initial colonial encounters were based on the idea of priority and of sameness/difference in which missionaries and agents of empire tried to make the Other more similar to them (civilized, human) through conquest/domination or protection. In other words, difference was understood relative to a continuous human spectrum, not a dichotomous nature/culture binary, although nothing guaranteed the indigenous a place on that human spectrum.

Consequently, in the sixteenth-century Spanish conception, the Other had different meanings, but two were dominant. In the first, the Other was a possible subject, but a helpless one that needed a process of transformation. This idea invoked to the humanist and Christian notions of egalitarianism and imperial politics of inclusion in which all human beings could become part of the realm (church or kingdom). Therefore, religion and conquest became allies in order to incorporate more souls and subjects. However, these imperfect human beings needed transformation to become perfect subjects. Equality did not arise immediately from the recognition of their humanity; instead regimens of exploitation were justified as a way of incorporating and eventually equalizing the Other through the civilizing process. Hard work or even slavery implied a purifying and improving process for the indigenous bodies and souls (Todorov 1984, Collier n.d.).

As for the second meaning, rather than a potential human subject, the Other could be possessed as an inhuman object (tool) or producer of objects (gold, silver and exotic products).

> It is held that since 1492 Europeans have projected an image of such people as somehow inferior, 'uncivilized,' and through the force of conquest have been able to impose this image of the conquered. The figure of Caliban has been held to epitomize this crushing portrait of contempt of New World aboriginals. (Taylor 1992:26)

This idea arose from the vision of the New World as pagan and uncivilized, in which the Other was considered to be an alterity (often an inhuman cannibal) in opposition to the "western self"—the "civilized" human (Todorov 1984, Collier n.d.). Spanish conceptions generated policies that imposed political and economic control over such objectified indigenous peoples, and this control was justified by the idea of indigenous incapacity to rule themselves as humans. Thus, the *Repartimiento de Indios*[3] and the *Encomienda*[4] allowed the conquerors and clerics direct control over indigenous peoples in order to some day transform them into human beings (Todorov 1984; Collier n.d.).

In Colombia, these two ideals of the Other (as essentially human or inhuman) found expression in two distinct colonial policies: the idea of the Other as an object justified the implementation of the E*ncomienda* and the conquerors' authority over Indians; and the idea of the Other as a human and potential subject justified the foundation of Indian Towns as centers for the imposition of Christian doctrine and the "saving" of indigenous souls. Through these towns, the Spanish crown hoped to reproduce European cities and humanity.

In Europe, royal courts constructed representations of the Other based on collections of objects and classifications of cultural expressions such as body painting and pottery. Eventually the physical presence of the Other as evidence of difference was required in these European courts, and the Spaniards took real bodies to Spain in order to provide that evidence. According to Collier (n.d), "beginning with Columbus, minerals, plants, and sometimes natives themselves accompanied [Spaniards] back to the European courts, often to form part of the encyclopedic collections."

Drawings and "collections of exotic indigenous peoples" reinforced these experiences as stereotypes. As stereotyped representations, these objects and persons could be "appropriated and transferred, [and] they provided the material out of which new stories about the world could be created" (Lutz and Collins 1993:35). Here, it is clear that indigenous peoples were making possible the construction of a Christian, European self that is a precursor to the modern western self.

In turn, these representations of the Other provoked a comparison between different cultures that usually prioritized the European as superior. In this way, notions of superiority and hierarchy were reinforced through representations that offered proof of the existence and 'inferiority" of the Other. Visual representations confirmed the inferiority of the Other because these images reproduced a consistent and therefore "real" western testimony and evidence of the Others' uncivilized world, which was reinforced as they traveled from place to place to feed the imagination of European consumers and to confirm their stereotypes. In short, these representations maintained and reproduced the inequalities between the incipient western self and the Other.

And so the European imaginary created stereotypes that marked and constrained the identity of the Other—as a subject to be subjected or as an object that could be possessed—and, at the same time, they constructed the European self as a subject capable of subjecting, possessing, valuing, observing, representing and judging the Other.

Although these early colonial representations constructed the Other inconsistent manners, they were complementary. The idea of the Other as cannibal or savage implied the necessity of conquest to eliminate this natural (an inhuman) difference, while the idea of the Other as sharing a common human identity or incomplete identity also implied that Spain should incorporate it through conquest and a systematic reduction of its difference in order to finish its "development" and "progress" and thus protect it from being unjustly excluded from humanity. Moreover, both ideas of the Other (despite their differences) have often functioned to confirm European superiority and to provide justification for processes of exclusion and violence that could be physical or symbolic and produce spaces of terror (Todorov 1984, Taussig 1987).

These two notions of the Other were closely related to conceptions of nature. This identification of indigenous peoples with nature has several causes, but I highlight two: the prevailing medieval notions of environmental determinism in which the tropics didn't allow the "bloom of culture," which implied that its inhabitants had a greater proximity to nature, making them almost animals; and the belief in moral recuperation and improvement of nature by means of exemplary cultural disciplines and activities (Borja 2002) that implied the necessity of moral and social reconstruction of America's indigenous peoples.

From the colonists' perspective, the asymmetrical power relationships that favored Europeans suggested a link between nature and the culture of inhabitants of a particular locale. These ideas reflected an environmental determinism in which nature created both bad habits and virtues. The tropics implied extravagance, extremes and passions that allowed uncontrolled feelings, while moderate weather, like that in the European empires, allowed people to balance and control their passions (Arnold 2000). And given the Christian assumption that human nature is corrupt, an environment that reduces human control over nature could only produce evil.

Renaissance conventions of narration also affected the representation of nature and humanity. Borja (2002) argues that narrative in the 16th century had the essential function of being a "teacher of life," that is to say, of narrating moral truths. The purpose of each narration "was that the reader identified bad habits and virtues embodied in the actions of the men" (Borja 2002:1). It also explained that these bad habits or virtues were caused by the influence of nature, which was also read and narrated according to the "rhetoric of figures." This rhetoric of narrative used the exemplum (example), the figure of the wonderful thing—representations of the reality associated with the supernatural and miraculous—to reinforce cultural virtues and highlight the bad habits associated with nature.

According to Borja (2002), Europeans "read" nature as unpredictable—lacking the ordered force of God—and capable of producing supernatural beings (wonderful animals, beasts). Likewise, nature (the jungle) was associated with savagery in opposition to the civilization and morality evident in the order of European cities. These images reinforced the idea of a sinful and corrupt nature that influenced the behavior of American natives. Borja (2002) states "According to that belief, man apprehended the basic characteristics from his geographical environment to the point of being identified morally with that natural environment" (14).

The physical environment determined social and moral behaviors and indicated the level of "evolution" of each social group, from savage to civilized. Therefore, it was necessary for the civilizing process to subject and to dominate terrible savage customs. In the same way, the image of "savage" nature provided Europeans the image of otherness thy needed to measure their own "civilized" character and to vindicate the virtues of the "conqueror" as demonstrated through their domination of nature (and "her" people) through violence. Once conquered, nature was given a new name—a new identity—that made it sacred and part of Christian territory rather than terra incognita.

At the end of the 18th century, the relationship between indigenous peoples and nature remained almost the same. Philosophers and historians from America (belonging to the intellectual Creole elite) declared that the difference between Europe and America resulted from the degree of civilization. They often represented American "Indians" as humanity's degenerate specie in contrast to the Rousseauistic ideal of "the noble savage" who lived in a paradise. These philosophers also employed conceptions of environmental determinism and cited the biological and environmental differences of the "new" continent (few and small species, geological alterations of nature, tropical climates and intensive humidity) to account for the differences of *Indians,* who were the enemies of all civilized work and progress. The Indians' state of environmentally-imposed lassitude and savagery therefore required the imposition of disciplinary processes to reorient them toward "civilization" (Brading 1994).

Among the images of *natives* that reigned in the moment of conquest and colony, it is important to highlight associations of *natives* with monkeys, especially humans with tails like monkeys, *homos silvestris,*[5] *cinocéfalos,*[6] and *orejones,*[7] among others, that reflected and reproduced Renaissance notions in America (Rojas-Mix 1992, Cabarcas 1994). These representations appeared in the first maps of America to show the spaces and natives that had to be conquered.

These notions are not far from later colonial views of nature as a resource requiring discovery and development: particularly the idea of a pristine nature awaiting industrial conquest (exploitation). The colonial notion of nature as terra incognita also encouraged the work of naturalists who generated inventories and counted plants and natural resources through botanical expeditions in which indigenous peoples' knowledge figured importantly, especially in the development of new medicines in the European court (Nieto 2001).

The Republican Culturalization

Indigenous peoples have had a variety of representations in the western imaginery depending on the particular period and interests of those producing the representations. However, the western imaginary is not simply the result of its own time and place; it is also the result of the struggle to represent encounters with otherness and difference and the power relationships those encounters initiate. Representations of the nonwestern Other in themselves inevitably produce relationships of power in their manner of production as well as their manner of consumption. And, as noted earlier, the historically derivative and cumulative tendency evident in western representations (stereotypes) of the indigenous Other creates an historical dimension that also transcends the immediate time and place of a given representations' production and exercises its own type of historical influence. As a result, contemporary representations of the *ecological native* and their effects cannot be understood without an understanding of the successive historical influences originating in the West that have become embedded within contemporary representations of indigenous peoples. The following sections will outline, from a Latin-American and Colombian perspective, the main historical influences that have contributed to the palimpsest-like quality of contemporary representations of the *ecological native*.

As the Spanish Americans *(criollos)* replaced the old colonial regime, they tried to follow the European ideal of the modern nation-state. The *Criollo* elite found inspiration not only in the American and French Revolutions' visions of social change and political organization, but also in doctrines of individual equality and free trade (Oquist 1980) that opposed the Spanish monarchy's authoritarianism and protectionist tributary economic relations and vindicated their republicanism, capitalist economic notions, and their desire for legal equality as national citizens.

Modernity quickly found acceptance among an elite that subscribed to the liberal European thought beginning to pervade the colonial situation, not only in Colombia but throughout Latin America. Some of these transformations began in the 18th century, when Latin America underwent clear attempts to reproduce European modernity. For example, Melo (1998) shows how in Colombia one of the principal strategies of modernization was the creation of academic institutions that helped to construct a national identity through a national science and a rethinking of Spanish institutions. In a similar way, Quijano (1995) shows how different Latin America countries, societies such as The Friends of the Nation constituded to promote "modern" ideals.

These processes continued in the 19th century. As Calderón (1995) notes, "a deformed nineteenth-century modernity was introduced to the continent by the old colonial elite, now republicans" (56). In this new republic, the white and liberal elite partially assimilated modernity, while the mixed and ethnic Others were still under a colonial system that contradicted the modern republican notion of citizenship. In fact, the hierarchical racial borders remained unchanged, although the dominant national policy, as it was in Europe, was that of the integration and assimilation of all minorities into a modern constitutional republic and citizenship.

Accordingly, the local elites assumed the idea of a society based on hierarchical racial conceptions (constructed from both the colonial and modern ideals) to impose their conceptions of the "only" way to become "civilized." "[The] colonial society at the end of the eighteenth century was thus differentiated ethnically between Peninsular Spanish, Spanish Americans [Criollos], Indians, Blacks, Mestizos, Mulatos, and Zambos. The Peninsular Spanish and the Spanish Americans were divided into classes on the basis of their economic activity, which was also the case with the rest of the pure and mixed ethnic types in this class-caste system of stratification" (Oquist 1980:30).

In this context, "human groups" located in the lower level of the social hierarchy had to be "improved" by rational means of political administration and economic production. European ideas of superiority and rationality imposed themselves at local levels through processes of modernization that employed principles of political and economic individualism, private property and the accumulation of surplus for reinvestment. The imposition of these political, economic and legal categories classified indigenous peoples as subjects of the state (citizens) with an obligation to fulfill the purposes and requirements of the state's laws, and so they defined the indigenous peoples' rights (or lack of them).

In this modern context, despite republican political ideas of equality, the indigenous Other was not considered a co-citizen, but rather an exotic stranger and a child. Moreover, indigenous peoples were regarded as incapable of rational thought; and so during the republican period, indigenous peoples were not allowed to participate in political processes or have equal civil rights or a minimum of socioeconomic services. And, of course, their cultural difference, when not represented as a lack or negativity, went unnoticed and unknown. In effect, a passive notion of citizenship was imposed upon indigenous peoples. Although the hegemonic discourse of republican modernity recognized them as part of nation building, it also designated them as incapable of governing themselves. This national republican discourse employed a notion of citizenship that implied not only inclusions but also exclusions, and so despite the formation of the new republic, ethnic and economic situations did not change for indigenous peoples. Consequently, this process of modernization characterized indigenous peoples as backward laborers needing "improvement" through discipline, missionization and pacification. Nevertheless, some indigenous peoples found ways to respond to this situation and establish negotiations with the new republican powers, although many others endured and adapted to the change or simply ignored the republic and continued with their cultural practices as before.

According to Quijano (1995), indigenous peoples' practices based on "reciprocity, solidarity, the control of chance, and the joyous intersubjectivity of collective work and communion with the world" also provided Europeans with models for their utopias and speculations regarding human nature (such as those of Rousseau). Nevertheless, the process of modernization in Colombia generally encouraged the local white elite to reproduce the European model rather than explore alternative indigenous visions of social order.

Although the anti-colonial struggles challenged the authoritarian colonial regime, the new nations did not possess the necessary socioeconomic and political conditions to achieve the "universal" ideals of nationhood, citizenship and democracy insofar as colonial attitudes and activities continued. In other words, Colombia has never realized modern ideals because the implementation of such goals requires a particular historical process, that was lived in Europe and that could not be reproduced in the same way in Latin America where it existed mainly as a desire. Likewise, specific socio-historical conditions in Latin America allowed a different interpretation and appropriation of these universal ideals. Thus, historically mixed realities characterized the region at the start of 20th century (Holston and Caldeira 1998, Caldeira 2001).

Despite such circumstances, Latin American elites, especially those in Colombia, pursued modernity and republicanism in accordance with a universalist vision of the Enlightenment and so their "expert knowledges" constructed a modern state and a set of political agendas that formed new relationships for its citizens of that state according to an apparently ahistorical standard of reason. This process displaced the old regimens of domination based on the survivals of colonial systems such as the *Hacienda* and the Catholic church and imposed new conceptions of the state based on the idea of a republican national identity and rationalized administration.

I do not want to develop here all the implications of the republican and modern ideals. However, I do want to highlight how modern conceptions of citizenship and a distinction between nature and culture, despite their differences from those of the colonial period, again put indigenous peoples under unequal relationships of power.

The modern duality of nature and culture helped *Criollos* clearly understand and construct their own cultural reality, yet it made their understanding of indigenous peoples and relationships to nature ambiguous. This modern or *Criollo* logic excluded indigenous peoples from full citizenship (humanity) either because they were considered part of nature or they had yet to develop sufficiently to become part of civilized life.

At the beginning of the 20th century, Latin America had its own interpretation of the modernist impetus. According to Calderón (1995), Latin America contributed "the intellectual elaboration of revolutionary nationalism and of national popular, or populist, movements" (57). In Latin America this project of modernity and its associated development programs did change social conditions and assimilate many of the excluded to modernity and citizenship, but it also caused many contradictions that provoked a crisis insofar as a lack of national economic resources to "develop" the countryside and cities resulted in weak institutions and corruption through bureaucratic malfeasance. Nonetheless, the national political goals remained the integration and assimilation of all "minorities" as citizens of a modern nation-state.

In the Colombian context since the 1950s the processes of modernization have been directed toward technological advance and efficient relations in production and commercial processes. Consequently, international experts (in economics, demography, education, agriculture, health and nutrition) have designed programs and determined local necessities in order to change and assimilate Colombia's "third world" inhabitants to the improved ways of "developed countries" (Escobar 1995).

The emergence of social movements in Latin America in the 1970s contested the state and changed notions of citizenship as well as democracy by challenging traditional authoritarian orders and by promoting democratic social processes that affected political representation. In this manner, social movements have acted to promote different social and political realities by reconfiguring modern national conceptions of identity, citizenship, political participation, nation, and development (see chapters 2, 3 and 4).

During the 1980s, social movements prompted new political relations with "minorities" in Colombia that included them as part of the national project. Since the 1980s, the state has pursed processes of constitutional revision resulting in the 1990s in the declaration of a New National Constitution. Thus, the New Political National Constitution (1991) which has opened political arenas for indigenous peoples and ethnic minorities that have enabled discussions about the process of modernization and its implications for indigenous and nonindigenous Colombians.

As a result, the Colombian state has in fact changed its processes and policies and included indigenous peoples in its discussions, which has reconfigured the state as an explicitly multicultural entity. Accordingly, the national state began to change its strategy of political participation to accommodate this new multiethnic conception of nationality (Gros 1998). Moreover, the crises of development and environmental change have also created a need to rethink processes of social and political representation and participation. So, too, these crises created a need to rethink and reconfigure representations of indigenous peoples in the ecological context of environmentalism.

The Environmental Re-naturalization

Since the 1970s, environmental changes around the world have attracted increasing concern and given rise to international policies that address the global ecological situation. In the case of Colombia, the first reponse was to form national parks as public spaces in order to conserve and preserve unique ecosystems. Subsequently, the national government instituted policies to regulate the use of natural resources such as water, fauna and flora (Código Nacional de Recursos Naturales 1974) and created new institutions to implement these policies (Inderena, Sistema de Parques Nacionales). These institutions enacted policies that promote the conservation of natural areas, the restriction of use in some natural or cultural territories, the control and regulation of natural resources, the reintroduction of endemic species in "natural territories" and the creation of zoos and places for the reproduction of wild species (botanic gardens, nurseries for wild animals, among others) (see chapter 4).

Colombian social movements have introduced not only new arenas for discussing rights, citizenship, identities and political participation, but also new arenas for discussing environmental problems and solutions (see chapter 2). For example, in response to environmental changes, social movements have promoted community reserves, collective protection of natural resources and social agreements to manage resources according to local practices. Among these social movements, indigenous peoples have assumed leadership in proposing different relations with the environment based on their notions of "development" and the relationship between humans and nonhumans. The confluence of indigenous actions, environmental changes, environmental organizations and ethnographic studies has produced national and international recognition of indigenous ecological knowledges. Consequently, their knowledges began to find greater acceptance and representation in the political discourse of Colombian and international governmental and nongovernmental organizations (see chapter 4).

These national and transnational processes eventually resulted in international recognition of indigenous peoples as environmentalists. This has mostly taken place since the end of the 1980s and beginning of the 1990s. In 1988, David Kopenawa (Yanomami) received recognition as an environmentalist from the United Nations. Since then, many other indigenous peoples from different parts of the world have been recognized as environmentalists for their contributions to conservation by diverse national and transnational governmental and nongovernmental organizations in the form of the Goldman Environmental Prize and the Right Livelihood Award. Initially, recognition tended to be granted to indigenous peoples principally from so-called "exotic" ecological regions (especially the rainforest), but nowadays such recognition is often granted to indigenous peoples from "nonexotic" parts of the world and for diverse activities related to conservation as well as sustainable development.

For example, the Isoceno people who live in Bolivia received recognition in 2002 from the Argentinean Foundation of Natural History Félix de Azara for their conservation of the environment and culture. In Spain, they also received the Bartolomé de las Casas prize for their ecological work in 2002. In the Bolivian context, in 2001, the Isoceno peoples' Management Plan for the National Park Gran Chaco Kaa Iya gained recognition from governmental institutions as a model for the National System of Protected Areas of that country.

In Colombia, official recognition of the relationship of indigenous peoples and the environment began in 1984 when COICA received the Alternative Nobel Prize, the "Right Livelihood Award," for its work and proposals that included environmental projects. After that, the Embera and Wounan peoples' grassroots organization (*Organización Regional Indígena Embera-Wounan-OREWA*) received the "Environmental National Distinction[9]" in 1995 for its achievements as an organization and environmentally-aware society. In 1997, eleven Embera researchers along with two biologists and one anthropologist were also awarded the environmental and development national prize "Ángel Escobar" for their participation in developing strategies for management of fauna employing Embera environmental practices in the Ensenada of Utría, Chocó. It is also important to note that this was the first time that indigenous peoples received recognition in the field of environmental scientific research.

In 1998, the Uwa people received the Goldman Environmental Prize.[10] The Uwa leader Berito Kuwaru'wa denounced nationally and internationally plans for oil exploration in their sacred territory. He asserted that suicide was preferable to accepting the destruction of their lands, and he emphasized their position by stating, "If we should die, the lights of the sky will be darkened." As a result, the Uwa received the support of environmentalists in the United States and countries such as Spain, Finland and Denmark. The defense of their territory and the world's environment has gained the Uwa people not only prizes, but international support and more territory.

In 1999, the Kogui people of the Sierra Nevada de Santa Marta, the most "traditional indigenous people"[11] in Colombia, received the international award for ecology from the Biopolitical International Organization (BIO).[12]

In 2001, the Union of *Yagé* Healers of the Colombian Amazon (*Unión de Médicos Indígenas Yageceros de la Amazonia Colombiana-UMIYAC*) received the Environmental National Distinction for its work linking traditional knowledge of medicinal plants and the use of the Yagé to sustainable development. They also received from the University of Valle an Honorable Mention in recognition of their medical practices, and recognition and homage from the Valle Department Government because of their ecological knowledge related to agricultural practices (Gutiérrez 2002).

Such recognition has made indigenous peoples' knowledges fundamental elements of environmental proposals and conservation programs in national and global sustainable development projects. Likewise, indigenous peoples' knowledges began to appear within national research policies as a form of knowledge that contributes to the maintenance of biological diversity.

The global economic and environmental crises have also generated more awareness of the indigenous systems of management of natural resources. Accordingly, the category of "primitive" and indigenous knowledge systems has been revaluated. In fact, the idea of "collective resource management" based on indigenous communal management of natural resources has taken on increasing importance in western environmental discourses. According to Moseley (1991), the practical significance of the "primitive" in the West arose because the goals and processes of industrialization and development programs have often put economic results and the public good represented by the environment in conflict with one another. Indigenous peoples' environmental knowledges have thus become part of the discourses of scholars, international agencies, NGOs, environmentalists and multinational corporations seeking means to make economic goals and activities consonant with those of a healthy environment.

As the preceding overview indicates, the representation of indigenous peoples as *ecological native* carries within it various and often contradictory influences and assumptions from a range of historical moments and contexts. As a predominantly western-controlled phenomenon, the representation of indigenous peoples is mainly a function of the West's needs and concerns, although the recent western shift toward environmentalism has created, in some cases, a more positive and dialectical relationship between westerners and indigenous peoples with respect to the production and consumption of representations of indigenous peoples. Nonetheless, the much older tendency to use representations of the indigenous to confirm what the West is not (inferior and undeveloped) and what the indigenous should be (more western and developed) continues.

Tennant (1994) has noted how this shift in the perception and representation of the indigenous peoples occured in international institutions and discourses, particularly how the representation of indigenous peoples has changed in those contexts from that of the ignoble primitive image to that of the noble primitive (see table 6.1). In particular, he notes how these two stereotypes have played important roles in the construction of the western self. In the first case, indigenous peoples appear as primitives, lacking civilization and living miserable existences. This image has characterized representations about the Other since colonial encounters. Tennant (1994) argues that during 1945–58, this representation was common among publications of the International Labour Organization, documents of the United Nations about decolonization and writings on indigenous issues from Latin America. The representation of indigenous peoples as primitives encouraged the imposition of expert knowledges and development programs designed to transform them into civilized people: "The ignoble primitive represents the antecedent state which the West has had to overcome, assimilate, and destroy in order to become modern" (1994:6).

In the second case, the representations of indigenous peoples in ecological discourses have contributed to the western image of the "noble primitive" (the good savage) who lives in a communal society and has a close and harmonic relationship with the environment. These are the noble primitives who fight economic development programs that have often destroyed their cultures and promise in turn to destroy modern societies. They also represent the West's vision of an environmental utopia.

According to Tennant (1994), this latter representation began in 1971 with the proliferation of documents from the United Nations, NGOs and environmental programs. In the imaginary of eco-tourism, for example, the return to the indigenous traditions appears as a hope for urban people. Diverse social movements (environmentalist, religious, and pacifist, among others) have also used this image to form critiques of the West and modern thought and alternatives to industrial societies.

This view has been reinforced in cinematographic productions such as "*Kayapó: Out of the Forest (1989),*" "*Blowpipes and Bulldozers (1988),*" *Fern Gully: The Last Rainforest (1992),*" "*Amazon Journal (1995),*" "*Tong Tana: The Lost Paradise (2001)*", or *The Shaman's Apprentice (2001).*"[13] Such films reaffirm the relationship between indigenous peoples and nature through images that portray indigenous peoples according to western expectations and stereotypes. These representations mystify and ignore "real" indigenous people and the various and complicated relationships that they maintain with their environmental, social and historical contexts. Nonetheless, many westerners continue to imagine the relationship between indigenous peoples and nature in terms of a preindustrial economic system based on autosubsistence. These representations in fact reflect the West's romantic conceptions of nature and its desire to deny its own various, complicated and contradictory relationships to the environment and indigenous peoples.

> *The Shaman's Apprentice*, like many other environmental films and the conservation programs they portray, has popular appeal because it offers a carefully crafted win-win vision of conservation and sustainable development, in which local participation and cultural survival are joined with the farsighted work of visionary scientist-activists. (Vivanco 2002:1202)

In short, indigenous peoples represent the western desire to return to a "primitive" mentality, a preindustrial lifestyle and an ecologically stable world. This representation has pervaded the mass media and thus, within the "global village," the representations of indigenous peoples (as homogenizing stereotypes) present them as wise ecological and communal healers. Moreover, the imaginary formed by the mass media represents the return to indigenous traditions as an antidote to the problems of "urban lives" and a criticism of the modern world. However, they also imply that expert knowledge and external intervention (lawyers, anthropologists or environmentalists) are necessary to protect the indigenous peoples from destruction and extinction, thus giving western peoples a dominant and heroic role in the story of the *ecological native* (a new myth of western absolution and salvation).

KOGUI ALLEGORY

The Kogui people occupy particular place in the Colombian and international imaginary that symbolizes pristine and pure knowledge in relation to nature (Uribe 1988, Milton 1996, Orrantia 2002). Within the Colombian context, particularly in the case of the SNSM, this symbolic role is evident in nonindigenous expressions of concern for the mountains' environmental deterioration and the indigenous peoples' role as protectors and saviors of the environment. This concern appears in local, national and international discourses that note the SNSM as a unique ecosystem, a great natural reservation, and the heritage of humanity. They also emphasize that it is the principal source of water for the region, and it also has the only pre-Columbian (the Lost City) site of indigenous cultures with significant urban development in the country.

However, serious problems of deterioration caused by colonizing activities and their inadequate productive practices, such as raising cattle and industrial monocultivation of banana and African palm, have damaged the SNSM's soils, as have the unlawful cultivation of marijuana and cocaine. These activities have caused indigenous peoples to move on to higher elevations, breaking with their management scheme of vertical zonification that involves utilization of both lowlands and highlands. According to environmental analysis, only 18% of the primary woods have been conserved, therefore there is a looming environmental emergency of water shortage and deforestation. Thus, saving the SNSM has become a local, national and international concern. However, it is useful to indicate here that this concern regarding the conservation of the SNSM aims also to assure its "development" under sustainable parameters.

It is in this context that the ecological identity of indigenous peoples of the SNSM emerges. Insofar as chapter 5 already addressed the process of construction of ecological identities, the focus here is on particular representations associated with those identities. In different magazines and newspapers the ecological identity of indigenous peoples of the SNSM is most often attributed to them by other agents (public representatives, journalists, NGOs, scholars, civil society) and has little to do with their own conceptions of their identities. Some journalists write their articles using idyllic expressions about the SNSM that locate indigenous peoples in the middle of an "exceptionally natural" setting where they live in "perfect harmony" with nature thanks to their "wise and calm spirit." They have promoted the idea that the indigenous peoples of the SNSM are nature's keepers who, through their ancient knowledge of their environment and ecologically sound practices, "make good use of the existing biodiversity without causing any deterioration to ecosystems" (Magazine from the Chamber of Commerce of Santa Marta, 05/92).

Western perceptions about the perfect communion between indigenous peoples and their territories, and their relation of respect and spiritual harmony with the environment also provide indigenous peoples with an important means to further their own struggle to keep their territories and conserve their resources. Some biologists and anthropologists argue that their ecological practices do in fact maintain a balance within the SNSM's ecosystem. They dismiss romantic notions and assert that "Experience has demonstrated to us that when the SNSM is under indigenous peoples' care it is regenerated, the river's flow increases and the woods are renewed" (*El Espectador,* March 12, 2001). Another author writes, "It is a community that has never stopped understanding that the balance of the world depends on our respect for the woods, rivers, lakes, wild animals and the sea" (Semana, October 18, 1998).

Nonetheless, even such factual views can be easily subsumed by the pristine image of purity and ancestral wisdom that often characterizes indigenous peoples of the SNSM in the western media. Such media exalt indigenous peoples' ancestral knowledge without addressing how it really works or how it may offer feasible alternatives to western plans to prevent environmental deterioration. Such vagueness does not stop them from assigning responsibility for managing the SNSM's ecosystems to indigenous peoples despite the fact that the indigenous peoples themselves state that "the damage is the responsibility of the white man who invaded our lands and exploited them without any control." One journalist noted this hypocrisy when he wrote "New settlers of the SNSM do not need an outside interpretation to realize that the enemy is themselves, and that like everyone, excepting indigenous peoples, they are using up totally their most valuable resource."[14] He also assigned a special responsibility to indigenous peoples when he stated that "Every sector has admitted the scope of the environmental crisis and their mutual responsibility, and have agreed to the urgency of enlarging indigenous peoples' *resguardos* to preserve the SNSM. They have learned that indigenous peoples are the only ones who for centuries have applied the concept that is in fashion in the Rio de Janeiro agreements: sustainable development." He also asserted that indigenous peoples have damaged their own environment only because they have been forced to exploit lands that according to their traditions were banned to occupation before the peasants and settlers invaded their territory.

Another ecological image of indigenous peoples of the SNSM appears clearly in an article written by Silvia Botero and Rosario Ortiz (who worked at the FPSNSM) in *Ecológica* (an environmental journal, No. 3, 1991, Bogotá). In this article, they offer a description of the critical environmental and social situation in the SNSM, and highlight the performance of the FPSNSM in programs developed for the conservation of biodiversity. They also refer to the Kogui people as the keepers of an oral tradition with pre-Spanish origin, and they note that their ancestors are the Tayrona peoples who left their expert environmental management printed on the landscape, until it was transformed by colonists coming from other zones of the country who did not value their cultural and ecological legacy. According to them, this colonization forced ancient inhabitants to move to higher parts near river headwaters and sacred zones where they had not been allowed to work previously. Indigenous peoples, regardless of the knowledge that they have of the environment's sustainable management, were thus forced to transform their practices.

Here, indigenous peoples are not blamed for the deterioration of the SNSM, rather peasants and new settlers take the blame for disrupting the environmental balance. Peasants deny this accusation and defend themselves by using the ecological discourses and representations of the ecological indigenous peoples to assert that they are also heirs of the Tayrona legacy,[15] as is apparent in a letter in which they rejected the national government's plan to spray the coca fields of the SNSM. They claim that the SNSM's deterioration is not due to peasant or farming activities, the majority of which are dedicated to growing their staple crops for their own support. They instead claim that it is the result of the national government's "unconcern and apathy in face of economic and social problems concerning the development of the region. It is also due to the state's neglect in which it has kept us throughout history. Our ancestors, the Tayrona, left us, according to their traces, an impressive ecological culture, but it was destroyed by the Spaniard ignorance" (Peasants letter, 1995). However, national nongovernmental and governmental institutions do not recognize peasants' claims as those of *ecological natives.*

Indigenous peoples' return to their ancestral land and the expansion of their *resguardos* to the sea is the officially sanctioned plan to conserve the SNSM. Different state institutions concur that the goal "is not only to return the possession of the *resguardos,* it is also to protect, preserve and develop in a sustainable manner the biodiversity, ecosystem and the cultural wealth of the Sierra Nevada" (*El Informador*, October 20, 1995). The director of the national institution of agrarian reform (Instituto Colombiano de la Reforma Agraria- Incora) said "The return of lands with access to the sea to indigenous peoples has a double purpose: one is to return them to their primitive owners, and the second is to favor their ecological conservation through an adequate management of the hydrographic basins of the sector in the hands of indigenous peoples" (*El Informador*, September 5, 1994). Julio Barragán, director of Indigenous Affairs of Magdalena states that "The return of these lands represents a social economic benefit for the communities because the market productivity and restoration of Don Diego and Palominio rivers may be carried out" (*El Tiempo*, July 25, 1994).

As is evident in the foregoing examples, governmental and nongovernmental institutions and various social and economic agents have used the image of the *ecological native* for their own purposes. Indigenous peoples in turn have reconstructed and used this image as a political strategy to recover their rights and autonomy in their territories and to manage their resources according to their own cultural traditions and practices. They use it to emphasize their claim that only they can achieve the "salvation" of the SNSM, and that only they can teach their destructive "younger brothers" how to live in harmony with nature:

> You have to give thanks [to us] for keeping the balance of the Mother (Sierra Nevada). This territory belonged to our ancestors, but they have been forced to go up to the mountains. Five hundred years after, the Government and the Constitution have recognized the right to claim our territory. We have to show our

> *younger brothers* what to do with nature. The Sierra is the heart of the universe; she had four keepers that took care of her: the sea, the wind, the earth and the fire. The *younger brothers* that came from the other side took the sea away from us, for this reason nature is sick. (Words pronounced by a Mama during the delivery of the territories with access to the sea. *El Tiempo*, July 26, 1994)

However, this political strategy also requires concessions:

> Indigenous peoples can administer and take care [of our territory] because it is our law, that is to say, to take care of nature because the future is there. If we destroy it we destroy ourselves. If the land is guaranteed to us, we will guarantee the water that will benefit the lower part [of the SNSM]. (Words of Adalberto Villafañe, after signing an agreement with the Governor of Magdalena for 60 million Colombia pesos to enlarge *resguardos. El Informador*, October 20, 1995).

This is a game of power in which we can clearly see that the purposeof non-indigenous social and political agents' acceptance of the indifenous peoples' ecological imge is the protection of their own economic interests (namely, the use of water resources). The governor of the Magdalena asserted "indigenous peoples will pay back this contribution with the maintenance of hydrographic basins and the ecosystem that is fundamental for the peoples' existence" (*El Informador*, October 20, 1995).

There are also larger political interests involved. In the case of the national government, there exists a desire to demonstrate to the international community that Colombia has indeed become a multicultural and pluriethnic nation as established in the CPC-91. The speech of César Gaviria during the delivery of land with access to the sea noted that "this expansion is given thanks to the pluriethnic and pluricultural society that is present today in Colombia. An example of this is the respect and admiration that every Colombian experiences through the convictions and beliefs of indigenous peoples and the support given to fulfill [their] expectations and aspirations, such as that the Government has provided through the Ecosierra Plan" (1997). For his part, the Director of the Indigenous Affairs of the Magdalena stated: "With this type of action, we are seeking a better relation between the state and indigenous peoples." A journalist confirmed this view by saying "The measure was taken in context of the new precepts of the Constitution of 91 in which the country is considered a pluralist republic. In this spirit, they also delivered the CPC-91 in the Ika language, which is their language."

Accordingly, the *ecological native* image is contradictory because it is simultaneously one of recognition and one of subordination. On one hand, the *elder brothers* (indigenous peoples) are the saviors of the world and the most clear and progressive in their thinking about the sustainable development of the SNSM, but on the other hand, their pristine condition (their image as good savages who live in an unchanged world) generates a parallel primitivist view of them. In the latter case, indigenous peoples are considered defenseless and passive beings who have to be protected and helped due to their ignorance, poverty and vulnerability. This problematic is evident in narrations such as the following:

> As from our point of view, they are extremely poor and we do not know why, especially if they have several acres of land planted, a farmyard with half a dozen sheep and a dozen cattle [. . .] They are poor because they lack services of health, education and everything that is fundamental for us. (*Cromos*, June 3, 1991)

Despite the recognition of indigenous peoples' ecological knowledge, some governmental officials assume that they should be guided and trained, tin a manner that, from a western point of view, will allow them to enter into modernity and development. For example, they should be taught to implement the organic cultivation of coffee for export to Japan (*El Espectador,* October 2, 2000). Journalists have also stated that indigenous peoples should be taught how to manage exotic plants for export (*Cromos*, June 3, 1991). Similarly, there are many workshops that seem to recognize indigenous peoples' knowledges, although they focus on the "environmental education" needed to implement sustainable development in the SNSM that will erase indigenous peoples' knowledges and practices by "improving" the quality of thought and life of indigenous communities, much as in colonial times.

In the end, there is a collision of competing interests regarding how to recognize and represent *ecological natives*. Although some do consider indigenous peoples' strong and active agents whose ecological wisdom is important for natural conservation and development; many still consider them vulnerable peoples who have to be provided with help and support. This would seem to be the view of the FPSNSM in its proposal to carry out training workshops among indigenous peoples' leaders as noted in an article that states "that the Foundation Prosierra intends to train indigenous peoples so they may participate actively in the recovery and conservation process of the SNSM" (FPSNSM 1997), This perspective clearly depicts indigenous peoples as passive and needing guidance in the development of their notions about nature and environmental management, which reproduces colonial discourses of "helping" the savage native.

IMPLICATIONS OF THE ECOLOGICAL NATIVE'S IMAGES

To live in harmony with nature has become a metaphor and an imperative in the global discourses that address the environmental crisis. However, this notion of harmony responds to western ideals of a lost and pristine Eden, which implies a nature that escapes cultural order. This perspective represents the "natives" as "wild" in opposition to "civilized" people of industrial societies. Likewise, this also implies a need for the intervention of an expert knowledge (anthropologists, environmental lawyers, environmentalists, biologists, and conservationists, among others) to protect indigenous peoples from destruction and extinction. This naturalized conception of the indigenous peoples suggests that they must be protected through conservation and preservation strategies the way rare biological species must be protected. This process also reflects the self-interest of these experts who want to perpetuate their own jobs, which are paid for by institutions whose interest is to demonstrate the dependence of the "natives" on the expert knoweldge they produce.

The mass media spreads and projects this image of the natives as integral components of an imperiled and exotic natural order rather than a group of fellow humans in a complex network of social and natural relationships. For example, in Colombia, regarding the relation of the Embera people to the Ministry of Environment, the journalists' narration did not focus on the Embera's problems but on how they paint their bodies and perform their rituals: "they sang and they played to Mother Earth and Father Sun, greeting the east, the north and the west where their gods are. These ceremonies were carried out amidst a jungle of asphalt and they smoked the pipe of peace amidst the gaze of military forces and the faces of terror of the executive women who had just left their offices when they saw the indigenous women topless and with body painting" (*El Espectador,* February 4, 2000). Such journalistic narratives also reinforce the image of the wise *ecological native,* especially when they highlight the "wisdom of our aboriginal cultures in their harmonic adaptation to the natural environment" (*El Colombiano*, July 4, 2000).

Journalists also frequently choose visual images that illustrate "exotic" *ecological natives* even when the image does not correspond to indigenous

Table 6.1. Environmental Discourses' Representations and their Implications for Indigenous Peoples

IMAGES OF ENVIRONMENTAL DISCOURSES	REPRESENTATIONS RELATED TO INDIGENOUS PEOPLES	IMPLICATIONS OF THESE IMAGES FOR INDIGENOUS PEOPLES
Mother Nature	• Wild or Nature's Children	→ They stand for biological diversity as species
	• Feminine Entities	→ They are represented under Western gender conceptions that imply relationships of power (domination or protection) over them
	• Global Heritage	→ Their territories are considered a global patrimony and they have to remain in an ideal past
	• Martyrs	→ They can sacrifice their lives for the environment
Sustainable development	• Sustainable Native	→ They can implement sustainable development in a natural way
	• Needing Training in Management of Natural Resources	→ They have to be trained and disciplined by sustainable development processes
	• Premodern	→ They are against progress and development
	• Hypermodern	→ They are not really indigenous peoples because they are not "traditional"

peoples in the accompanying text. An article about indigenous peoples from Tolima uses an image of the Kogui, Uwa or Embera peoples because their dresses or body painting mark difference in a more evocative way or more in accordance with the expectations of their audience. In other cases the strategy is to simply assert the presence of "tradition" in term of dress or body display. Images of *ecological natives* may thus quickly develop into categorical homogenizing stereotypes for indigenous people in general. Due to the popularity of these stereotypes among western consumers of the media, indigenous peoples are often imagined in a noble state insofar as "Northern cultural preservationists wish to see exotic peoples preserved as idealized, superior and untainted by the market economy" (Alcorn 1994:8). These romantic idealizations require such categorical images, specifically images of societies with territories far from modern centers occupied by peoples, with an ecological identity and social institutions that allow communal cohesion.

Images that represent the *ecological native* are in fact multiple. However, in global environmental discourses two main ideas dominate. As we have seen in chapter 4: one related to biocentrism (monism) and the other modernity (dualism). Nevertheless, they are complementary and each needs the other in order to generate discussion, contradiction, opposition and agreement. These two ideas have also generated two distinct representations that are recurrent in environmental discourses: Mother Nature and "the indigenous other" as part of sustainable development processes (see table 6.1). westerners often represent indigenous peoples as Mother Nature's children, guardians of nature or personifications of the ideals of nurturing femininity. However, at the same time, protection of Mother Nature implies development workshops, forums and environmental education for indigenous and peasants communities in order to teach them how to guard the Mother through sustainable practices that allow westerners to pursue and realize their own interests. This idea of Mother Nature is closer to western conceptions than those of indigenous peoples, even though western environmental discourses often invoke indigenous peoples' conceptions of Mother Earth. The basic difference is that representations generated from the development perspective assume the existence of an alienation of indigenous people from nature that must be overcome, which is a view that does not necessarily exist in indigenous conceptions of their relationships to Mother Earth.

Mother Nature

One of the principal images that sustains environmentalist discourse is that of pristine nature represented as "Mother Nature" or "Mother Earth."[16] She is conceived as the image of life on which we all depend, from whom we feed and who guarantees our survival. She is the mother that gives life. She is abundant, fertile and exuberant. All the beings that inhabit her are parts and results of her creation. Human beings live in her lap as in a garden, which humans should not disrupt or modify. The urgency to preserve her has become a global priority. Western alienation from Mother Nature is frequently associated with a lost paradise that all humans desire to recover. This image of nature has inspired the preservationist programs and has helped to motivate national policies such as the formation of uninhabited protected areas in order to conserve her natural state, which may be seen as a rather paradoxical way to overcome alienation from nature and recover humanity's relationship to the environment.

The image of Mother Nature that environmentalist discourses use is related to indigenous peoples' conceptions and worldviews insofar as it represents the Earth as a vital principle of life. However, in environmental discourses indigenous peoples' relationships with the Earth are not very well explained and the image has become an icon without cultural context. Preservationist programs thus use Mother Nature to represent a pristine nature that fits modern western conceptions, but is foreign to those whose lives are more immediately dependent on "her" well-being.

In Western thinking, Mother Nature is a natural entity that has children, and among her favorites are indigenous peoples. This is because western environmental discourses attribute their "natural" spirituality and their harmonic relationships with Her to being Her creatures rather than humans. Therefore, protection of Her children implies programs for saving natural species in the process of extinction. Although this image of Mother Nature helps mystify Her and protect Her from environmental destruction, such a view has negative implications for indigenous peoples.

The Other as Wild or Nature's Children

It is common to find in publications about biological diversity a close association between indigenous peoples and biological species. Although one can imagine that images of indigenous peoples exist to recognize them as protectors of biodiversity, it is important to point out how these images often appear without historical, social and cultural context, and how the accompanying texts do not make reference to indigenous knowledges or ecological practices. These representations also frequently feature indigenous peoples in "traditional" dress that they do not use, or they are old pictures or pictures taken from old books (as we see frequently in the colonial process of representing the Other). However, it is important to highlight that because of this new idea of the *ecological native,* some indigenous peoples have begun to wear body paintings and feathers, even if they did not use them before, in order to mark their difference and because these images help them gain more support from national and international institutions.

The images of book covers,[17] posters and pamphlets used by environmentalists related to biological diversity almost always use indigenous peoples' images. These images present a visual continuity between the selected biological species and the images of indigenous peoples through color and representation of corporal decoration or use of "natural elements" such as feathers. Most of these images highlight a graphic similarity with animal species. As a result, indigenous peoples stand for biological rather than human diversity and so indigenous peoples often appear naked and without 'cultural' marks in order to emphasize their "natural" condition. These associations have visual continuity with colonial processes and generate at a glance an "animalization" of indigenous peoples that removes them from their historical and cultural situations as humans.

In global environmental discourses, indigenous peoples are located as "wild" and in opposition to people in agricultural or industrial societies. In the graphic productions of environmentalists about biodiversity, other cultures besides those of indigenous peoples are not commonly represented (for example, peasants or urban peoples who also produce biodiversity), and this is so mainly because their rural and urban representations don't generate that exoticization and hybridization with biological species that "natural" indigenous people "should" represent.

The Other as Feminine Entity

In addition, the western imaginary represents the environment in general and "nature" in particular according conceptions of gender. Nature is seen as a female (Mother Nature) that, according to western gender conceptions, implies relationships of power (domination or protection). Moreover, it is also assumed, by extension, that a feminine natural sensibility and spirituality can be found in the nature-based spiritual traditions of indigenous peoples.

In this sense indigenous peoples are thought to be, more than ever, part of nature, and like nature they are feminized. In this association of the feminine with nature, nature is always conquered or possessed (since colonial times), therefore, by extension, indigenous peoples become part of this idea. Moreover, western dichotomies that assert the dominance or even opposition of men and culture over women and nature are imposed upon indigenous peoples' conceptions of gender and nature with little consideration of the inappropriateness or harmfulness these patriarchal relations of power and
inequality may have upon indigenous beliefs[18] (Sturgeon 1997). These associations (mother=nature) can be seen frequently in representations of naked indigenous mothers with children that imply their naturalization through biological reproduction.

The Other as Global Heritage

Declaring the areas of greatest biodiversity in the world a universal patrimony or global heritage (and most of them are in the "third world" countries and belong to indigenous peoples, peasants or local communities) implies that their inhabitants have the responsibility of conserving, protecting and taking care of that biodiversity for the rest of humanity. Nonetheless, local inhabitants' expectations of development are rarely considered. Implicit in this idea of nature is another of a global form of property (in a symbolic sense and in a very practical and real sense) that belongs to all humans.

Nature can also be exploited and used because it is in a position of inferiority in relation to human necessities. This is a variation of nature under a pristine vision, because, in this case, nature is viewed like a garden or a patch of wilderness that should satisfy the collective environmental rather than individual industrial dreams of human beings. Nature in this perspective retains its link to the biblical conception in which God created paradise and said to "man" that all that was there belonged to him, except for the tree of knowledge, after which the expulsion of men from Eden unchained the process of transformation of the environment. To restore nature from the western perspective is thus (paradoxically) to recover the global paradise by asserting humanity's domination over nature.

Along with the idea of global heritage emerges the idea that "we" (all humans) have to protect the "last rain forests." Deforestation or desertification is now a global problem, and these forests are said to belong to "humanity," although not all are equally responsible for their salvation. Members of governmental and nongovernmental institutions often take for granted that indigenous peoples have to assume a great role in the historical and "natural" task of saving these forests and, therefore, the earth for the rest of us. Indigenous peoples have to situate their activities and proposals within this discursive context because indigenous people's territories are considered a global patrimony.

The Other as Martyr

Another image that produces the articulation of indigenous peoples and Mother Earth is that of stewardship. Indigenous peoples are the custodians and defenders of "her" diversity. Indigenous peoples become heroes or martyrs in defense of nature, which enlarges and ennobles them in the event of being "sacrificed" to protect her (their territories). These images also imply that indigenous peoples should have certain characteristics and behaviors in relation to the environment: they cannot be modern; they must not want to be modern; they cannot have technology; and they cannot opt for forms of life that contradict western representations of *ecological natives*. These naturalized images of *ecological natives* without history are those that many environmentalists and members of NGOs desire and reproduce (the symbolic eco-heroes).

For example, in relation to martyrdom of the Uwa peoples Gracia Francescato, a speaker of the Italian green party, said, "we have to stop the collective death of a whole people because of economic interests" (*El Tiempo*, June 27, 2000). In a similar way, the Argentinean Nobel Peace Prize winner, Adolfo Pérez Esquivel, in Rome and in front of the members of the international environmental organization Lelio Basso, gave his support for the Uwa's cause and said "I give my support to the Uwa people in their struggles. They are a symbolic example to the entire continent." (*El Tiempo*, June 27, 2000).

Indigenous peoples become the donors of life to humanity. In the vision of many environmentalists, indigenous peoples have to assume the historical task of saving the planet or Mother Earth by maintaining and perpetuating "traditional" ecological systems under the ecologically romantic vision of the "noble savage." However, most transnational environmental networks and movements have no realistic understanding of indigenous peoples' knowledges about nature and their perspectives on development. Moreover, within global ecological discourse, extinction and pollution appear as the result of "human" activities that are rarely specified as primarily those of western peoples. Although this implies that western peoples have a special obligation to find solutions to the global environmental problem, western discussions consistently refer to the obligation of those who still have a harmonic relationship with nature to provide those solutions. Therefore, indigenous peoples should save the planet (Mother Nature) for the rest of us despite the fact that we have done the most to harm it.

Indigenous peoples are now called upon to give their knowledge (which was previously unknown to western science) and genetic resources to humanity as expressions of solidarity with the rest of us because their knowledge and genetic material are part of nature and the world's patrimony. Indigenous peoples thus have the historical responsibility to maintain "life" (genetic resources) and help to reproduce humanity and other species. In this sense, they have the historical responsibility to protect their territories and maintain biodiversity without changing their cultural practices, as if in a photographic freeze frame.

The previous images associated with Mother Nature have introduced *ecological natives* to a new eco-governmentality and new eco-disciplines not only with respect to their territories but also their bodies whose genetic material has also become an object of western interest. This new eco-governmentality seems to have as an objective the maintenance of standards of life in industrial societies whose people don't want to change capitalist patterns of production and consumption. But it is not clear who will share the benefits of these newly discovered treasures of biodiversity, how they will be shared or whether they will really contribute to the preservation of the global environment.

All these representations of *ecological natives* associated with Mother Nature demonstrate how indigenous peoples are thought to be, now more than ever, part of nature. Likewise, images of the "traditional native" or of the "good savage" coincide with modern relationships with nature: colonization (human control of nature through domestication-civilization processes), industrialization (the technical transformation of nature into a resource) and protection (care of species threatened with extinction). Environmental discourse represent indigenous peoples as an Other who is external to civilization and therefore part of the natural environment. The production and consumption of these images of indigenous peoples reaffirm and maintain neocolonial or even biocolonial relations of power over their lives. In short, representations associated with a pristine nature are imposed identities that help to continue the stereotyping of indigenous peoples as the "exotic other" or the "good savage" for the benefit of the West as they did in colonial times.

Sustainable Development

The idea of an imminent environmental disaster, although it has scientific bases and evident consequences, sometimes is an apocalyptic vision. Global environmental problems like global-warming, industrial contamination, emanation of toxic gases, deforestation, extinction of species and the use of fossil fuels have been the primary motives for the creation of global environmental law. However, the manner of addressing these problems seems to predict the arrival of a chaotic time of the apocalyptic "beast" that could lead to the extinction of human species. Biologists, as if on a "crusade," are going to the last rain forest to count and save all natural resources before this apocalypse. There is a desperate call "to save nature" but there is little corresponding passion to state clearly what it must be saved from or to question whether western consumption patterns and conceptions of progress may in fact be the root of the problem. Therefore, western programs and projects in which conservation and management of natural resources are the fundamental axes often emphasize sustainable development as the only viable alternative. Most of the time, advocates of these programs propose environmental redemption and salvation without awareness of the potentially "apocalyptic" impact their proposed "solutions" may have on indigenous peoples or their territories. They often demonstrate a sharp awareness that there are some natural resources that have to be protected and conserved as biodiversity for humanity; but at the same time others essential to western lifestyles such as oil can be exploited despite the consequences for biodiversity and humanity.

Some sustainable development programs have designated specific activities (through global forums and workshops for the protection of natural resources) as means to sustain the global environment: the creation of sinks CO2 in countries with high biodiversity (which are mainly "third-world" countries) so that their forests will absorb the contamination emitted by industrial countries ("first-world" countries); reforestation; sustainable exploitation of natural resources; natural-resource management programs; and the marketing of environmental services that, of course, represent exceptional business opportunities, just to name a few. In a more spiritual sense, the environmental crisis promotes desires for more respectful and harmonic relations with nature among westerners who hope to follow the path of indigenous peoples back to a western vision of ecological paradise. Two basic ideas are thus present in the western approaches to environmental problems: a global economic strategy and the assimilation of indigenous peoples' spiritual conceptions to western ends.

The West needs indigenous peoples because only their "natural" products and beautiful territories (rich in biological diversity) can save modern societies from their spiritual and environmental crisis. Accordingly, the indigenous peoples must give inhabitants of industrial societies the opportunity to save themselves by consuming natural paper, organic coffee or "eco-products, "visiting beautiful settings through "eco-tourism," using alternative systems of healing or medicinal plants, and displacing their contamination to indigenous peoples' forests in order to clean "their" air.

When indigenous peoples are not responsible for preserving pristine nature and rare species under museum-like conditions, they are responsible for protecting the resources necessary to economic development (sustainable growth) that were previously of little value to western economies (genetic materials, for example). In short, the planet needs indigenous peoples to be part of sustainable development in order to save the developed areas of the planet. This requires indigenous peoples to be the guardians of nature while at the same time they are encouraged to participate within the global markets as new producers of raw materials, finished products and media images. Either way, the West views indigenous peoples as the only ones who can make the sacrifices needed to save "our" Mother Nature.

As Escobar (1995) shows in his analysis of development in Latin America, and Ferguson (1990) in his analysis of the political system in Africa, such contradictions were basic to modern conceptions of development and now are reproduced in models of sustainable development, with the difference that now indigenous peoples and local communities do not necessarily need to progress or develop. In models of economic development the Other had to change, but with sustainable development the Other often has to remain the same and not change, despite the radically different relation to the world such development demands.

Indigenous peoples have become agents within the global political arenas and participants in the formulation of global policies for sustainable development that value indigenous knowledge. In this sense, the environmental crisis appears to have rebounded to the benefit of indigenous peoples in the form of an apparent autonomy to manage their own territories. This is illusory, however, to the extent that they are subjected to inclusion in projects of sustainable development that promote a western logic and standards of valuation alien to their visions. There is a subtle method of domination—"you are those who know, but we need to teach you the way to value your knowledge and use it"—which continues the paternalistic colonial logic through a combination of the good savage with the one that should be civilized.

Representations of indigenous peoples also constitute a condensation of the values underlying the western idea of sustainable development. After Rio-92, sustainable development became not only a theoretical salvation but also as a global truth that had to be practically implemented. Therefore, the *ecological native* becomes an icon of this neutral idea that rises above political debate. Anyone opposed to sustainability denies a "natural" truth, and s/he has to be educated or condemned.

In discussions of managing natural resources and biodiversity according to sustainable development, the representations of indigenous peoples often have oscillated between stereotypes of the noble savage (when they protect the environment) and the ecological destroyer (when they consume or sell their resources[19]). In several studies, specifically those related to hunting, their image is double: one states that indigenous peoples conserve and protect animals, but at the same time the other image blames indigenous peoples for reductions in animal populations. In any case, the West values them according to its own indicators of sustainability, ignoring other global factors and the cultural complexity of indigenous peoples. I want to highlight the following variations of the West's representations of the indigenous Other before closing this chapter: the sustainable native, the native needing training in management of natural resources, the premodern native and the hypermodern native.

The Other as Sustainable Native20

Indigenous peoples have recently been included in environmental projects in which they help by participating in the design of various development programs. For example, in Bolivia, the Isoceno people have participated in the evaluation of the impact, design and construction of the Bolivia-Brazil oil pipeline with the representatives of the Viceministry of Hydrocarbons, Gas Trans-Bolivian and the Captaincy of the High and Low Izozog.[21]

Such sustainable-development discourses also depict indigenous peoples as having an interest in natural-resource management policies proposed by the state or international and national corporations. This perspective assumes that local people know their reality more clearly, and therefore local people are most capable of managing natural resources through local forms of organization and thought. Under this rubric occurs the idea that involving local people in economic development is the best way to protect biodiversity (Brosius et al. 1998).

Some anthropological and biological studies have shown how indigenous peoples have already been managing their resources in a sustainable way and already have the concept of caring for nature and biodiversity. From the western perspective, such indigenous peoples are seen as "natural" examples of sustainable management of natural resources, and western expert knowledge somehow finds in them a mirror image of its own assumptions about development and sustainability.

In general, western environmental projects seek the participation of local people who, in turn, give their knowledge, experience and hard work (it is cheaper to work with local people who know the context rather than bringing people from outside). When local people participate, it seems that sustainable development has become a collective and "indigenized" goal. Therefore, most of the oil corporations have programs that include indigenous participation and promote "their" worldviews in programs of environmental "education" that ultimately legitimize western assumptions and goals: economic growth and increased consumption.

The Other as Needing Training in the Management of Natural Resources

Other western representations depict local people as obstacles to conservationist conceptions of sustainable use. The indigenous appear as thoughtless consumers who maximize the use of natural resources because they do not think of the future. These images respond to the idea of the necessity of training local people because their logic doesn't correspond to the "true" logic of conservation (Brosius et al. 1998).

Most of the oil companies are now "green" companies. For example, the Shell Oil Corporation has natural-resource conservation projects aimed at protecting and conserving the ecosystem of the Gulf of Mexico (http://www.countonshell.com). Therefore, there is no reason for people to oppose sustainable development. On the contrary, peoples in opposition to development are against progress and against the salvation of nature, therefore they have to be "civilized" or 'eco-civilized" so that they do not waste or abuse the natural resources in their territories. Worse yet, they are against their own best cultural interests when they reject such western help.

The Other as Premodern

Even though indigenous peoples supposedly represent sustainable development in the West's "green" representations, they can still be represented as the enemies of progress. In the Colombian case, the processes that faced the Embera-katio and Uwa peoples demonstrate a contradiction involving sustainable development's economics and indigenous environmental values. Despite all their own mobilizations and the international and national support that they received, the Urra dam was built on their territory and Ecopetrol (a national oil corporation) continued with oil exploration near the Uwa people's territories.

Indigenous peoples, although frequently recognized for their ecological wisdom, may also be depicted as obstructing national and global development. Some institutions of the Colombian government continue to reduce the implications of development programs in indigenous peoples' territories to simple economic terms. National newspapers represented the actions of the Embera-katio people as an economic loss for the nation. The press argued that the protests of the Embera-katio had causes delay in the construction of the dam and a loss of 80 million dollars. A group of senators from the Caribbean coast of Colombia openly accused the Embera-katio of making losses to the state of $38 billion pesos through delays in filling the reservoir.

In the context of Colombia's national policy for sustainable development, the building of the hydroelectric dam can only bring benefits, because it will decrease the risks of energy shortages in the Caribbean region. In the words of the Minister of Mines, Carlos Caballero Argaez, "the hydroelectric [project] will reinforce the national [electric] system [. . .] Urrá will also be a motor for economic and social development of the region, since 6% of the energy sales will be dedicated to public benefit " (*El Tiempo*, February 16, 2000). The president of Urrá, Alfredo Solano Berrio, made a similar statement when speaking of "the social benefits" of the project in spite of its high cost ($800 million dollars), "because it will produce yields for energy generation and it will be an alternative to the thermoelectric production of the Atlantic coast, which will control the floods of the River Sinú and it will allow the agricultural development of 2000 fertile hectares that remain waterlogged during half of the year" (*El Colombiano*, September 5,1999).

Similar discussions surrounded the activities of the OXY in the Uwa people's territory. Governmental concerns centered on what Oxy's oil exploration projects would mean for the finances of the nation. In this case, there was a conceptual difference and struggle over meanings between the indigenous peoples and western participants with regard to management, the environment and natural resources. For the Uwa people, oil exploration was forbidden because the petroleum is the mother earth's blood. For them the value of economic projects cannot be more than that of man's life and that of the planet earth. The Uwa people tried to stop the project. They made judicial petitions and bought lands around their *resguardos* thanks to the economic support that different NGOs in Europe and United States offered them and with resources of the Prince of Asturias Prize they had won three years before in Spain.

However, for a national government under the mandate of development, petroleum is a prized source of economic progress for the nation. The national government and Ecopetrol alleged that the exploration was taking place outside the indigenous peoples' *resguardo* without affecting the interests of that community, and for this reason there was no consultation with them and therefore no reason to stop the work of OXY. They also said that stopping the exploration would mean "the municipalities and departments would stop receiving the 1.4 billion pesos of royalties and the government would not receive from Ecopetrol the two billion pesos to finance schools and hospitals" (*El Tiempo*, March 1, 2000). They also stated that besides affecting the oil self-sufficiency of the nation in petroleum, it would lead to the need to import crude oil, and this, according to the minister of Mines, "will be a serious disaster for a national economy that has been in recession for the last 60 years" (*El Tiempo*, October 11, 1999).

In spite of the government's hopes that economic development would improve oil exploration, OXY announced in July of 2002 the suspension of exploration because there was no petroleum. This means, at this moment, a victory for Uwa people in their struggle to protect their ancestral territory. However, the national company Ecopetrol will continue the exploration process again in the future, and in that case it will be a national and governmental institution against which the Uwa people may not be able to generate enough international support.

Sustainable development implies the recognition and the negation of indigenous difference simultaneously, and this discourse is repeated in each new plan or in each new development strategy. An example of this is the case of the SNSM, where the Colombian government simultaneously promotes the conservation of pristine nature and the application of sustainable development's policies. As Escobar points out (1999), development is exactly the mechanism through which indigenous difference will be eliminated.

The Other as Hypermodern

Within the idea of sustainable development there are indigenous peoples who do not fit the model because they are neither *ecological natives* nor people needing development.[22] Rather, these are indigenous peoples who live in cities, have businesses and do not use feathers or body painting, which is to say they are not "exotic." They have different visions about their territories and products. They market not only their images (there are many web pages through which it is possible to buy traditional instruments, dresses, food or music) but also their territories, as in the case of California where they have constructed lucrative casinos on their lands. However, to many westerners these are not "real" indigenous peoples, and so the western stereotypes of the *ecological native* affect them by imposing a frozen identity that creates obstacles to the hypermodern indigenous peoples' efforts to find and develop bases of solidarity with so-called *ecological natives.*

FINAL REMARKS: *ECOLOGICAL NATIVES*' IMAGES, NEW COLONIAL REPRESENTATIONS?

The *ecological native* implies ideal images which can legitimize indigenous peoples' knowledges and ecological practices (or western versions of them). Such western recognition, however, often comes at the price of subordinating indigenous peoples' political struggles to western visions of global environmental development. The Uwa people, for example, have had to adopt some of the norms and values of the national and transnational environmental organizations that have helped them resist the development plans of a multinational oil corporation. However, other indigenous peoples who willingly sell oil, (the Wayuu people of Colombia for example) or want to market their cultural heritage run the risk of losing western recognition and support insofar as they no longer conform to the stereotype of the *ecological native*.

Western actors have manipulated representations of the *ecological native* to support their own values and agendas, and these include governmental and nongovernmental organizations as well as national and international environmentalist movements. Indigenous peoples thus find themselves in an ambiguous relationship to representations of the *ecological native*. At times, it is a mask placed upon them to serve nonindigenous peoples' interests; and at other times it is a strategy that they have appropriated and refashioned and placed upon themselves to serve their own interests.

However, as noted above, western representations of the *ecological native* reflect the variety of often contradictory demands that the western imaginary places upon indigenous peoples who are at once nature's wild children, feminine entities, a global heritage, eco-martyrs, sustainable natives, hypermodernists, postmodernists or premodernists—all of whom either need training and guidance or guardianship with regards to the management of their natural resources.

On the one extreme of the western imaginary, the *ecological natives* are superior in their difference (ecological wisdom) and become models for western imitation, and on the other they are inferior (undeveloped) and need to be made similar to the dominant, developed and more "intelligent" western peoples. In both of these cases, however, indigenous peoples become the objects of western knowledge, interests and management. They are either too wise and pure to face modernity, manage their territories and preserve their cultures without western guardianship, or they are so ignorant and susceptible to the temptation of western consumer society and its benefits that their territories must be protected from them for the good of humanity. Accordingly, the environmentalism that emerged in the 1970s placed indigenous peoples at the center of the debate regarding how to preserve nature, especially areas of great biodiversity, from human harm. To the extent that indigenous peoples are not considered parts of the nature to be preserved (but instead potentially "harmful" humans), they became objects of critical scrutiny for those who conceive of preserving nature as the defense of a global patrimony and, therefore, a moral imperative. As Alcorn (1994) states regarding indigenous peoples as conservationists, "the question reverberates with echoes of the early colonial era debate (then in the name of 'manifest destiny,' now 'global heritage') over whether serving the 'greater good' gives the Europeans the right to ignore indigenous peoples' human rights and their preexisting tenurial rights over their resources" (1994:7). In either case, indigenous peoples' authorities are judged unable to adequately protect or govern the territories that have become humanity's property.

The idea of biodiversity has also placed the nation state in conflict with global interests over questions regarding the ownership, development and sale of valuable resources found in indigenous territories. Whereas global interests often promote "humanity's" interest in conservation or sustainable development, the nation state promotes its interests in economic development and political sovereignty over its territories. Although both give some recognition to the indigenous peoples on the territories in dispute, the underlying assumption is often that indigenous peoples are, at best, junior partners in the negotiation of such issues or, at worst, spectators or impediments to their resolution.

The politicization of biodiversity in multiple forms—as an economic resource or an environmental patrimony—contradicts the relationships that indigenous peoples have with nature. Indigenous peoples' knowledges of nature are complex cultural constructions that involve not only objects but also processes that are deeply spiritual, historical and politically relational. This complex reality is often ignored in the West's view of its political relations with indigenous peoples (which are often presented as ahistorical and apolitical techniques of management based on a transcendent reason). Consequently, the West replaces real relations with indigenous peoples with imaginary relations: to a grateful undeveloped ecological native who desires and thus legitimizes the western drive to impose sustainable development plans; to a wise *ecological native* who can give western peoples the environmental knowledge and spiritual means they need to correct and absolve themselves of the ecological devastation they have caused; to a destructive *ecological native* who threatens the world's patrimony. However, unlike imaginary indigenous interlocutors, real indigenous peoples are less likely to express the gratitude, grant the knowledge and absolution or confirm fears in the ways nonindigenous peoples desire.

Nonetheless, western peoples continue their ritual invocations of the *ecological native*. The physical presence of this native is required at concerts benefiting the environment (Sting sings beside Raoni, who the European parliament had just given money for a campaign supporting indigenous peoples). The environmentalists conjure indigenous "specimens" and "green bodies" to demonstrate their own spiritual purity and superiority vis a vis the West. In the 1980s, David Kopenawa's presence was invoked in international contexts to legitimize the relation between ancestral wisdom and biodiversity. So, too, in Colombia the Kogui *Mamas* provide all the illustrious visitors to the country (presidents, writers, and artists, among others) an opportunity to burnish their environmental credentials with the *aseguranzas*[23] they receive when touring the Lost City. Indigenous peoples thus continue to serve the West, not only as exotic others or resources for eco-capitalism, but also as spiritual models and resources for the construction of the postindustrial western self.

The western environmental imaginary imposes stereotypes that mark and limit the identity of this Other and, at the same time, builds ecological identities for western individuals that allow them to possess, observe, remember, judge and transform the *ecological native*. The wise *ecological native* cannot realize her global destiny without being subsumed and completed within the western imaginary, just as the wild or destructive *ecological native* cannot fulfill hers without the protection and guidance of western experts. In either case, completion occurs according to western paradigms and representations of ecological consciousness, economic development and environmental conservation.

In colonial times the various disciplines and processes used to transform indigenous souls and bodies shared one common feature: the colonizers had the right and responsibility to exercise such transformational power. Similarly, production and consumption of green cultural images of the Other (books, movies, museums, and anthropological studies, among others) reinforce and maintain relationships of power and transformation over the indigenous Other. In response, indigenous movements have begun to contest and modify conferred western representations and identities by producing their own representations in videos, texts, pictures and objects (Ginsburg 1994, Tennant 1994, Gupta 1998, Coombe 1998). In this manner, indigenous peoples have used conferred identities as a representational strategy to assert their difference (Gupta 1998). Turner (2002) also notes that "for indigenous peoples struggling to redefine the terms of their relations to the national and transnational systems in which they are embedded, representational media such as video have been useful primarily as means of connection in time and space" (244).

Although western conceptions of the Other constrain indigenous identity, the dynamic relations indigenous peoples maintain with the state and its nonindigenous citizens give them an opportunity to reconstruct their identities according to their own visions. And although they have inherited conferred identities, indigenous peoples' own processes of identity construction have established dialogic relations with those conferred identities that have allowed them to appropriate them for their own uses or for sharing with others on indigenous terms. Certainly, we can say that indigenous peoples are experiencing a dialectical encounter with western representations in which they have taken an active and assertive role.

Indigenous peoples understand that, in a postmodern world, control over the means and content of representation has an importance approaching that of the control they seek over their territories and cultural traditions. Accordingly, they have begun the long, difficult process of contesting the image of indigenous peoples as *ecological natives*—an image that has often been used in international arenas as a means to replace political negotiation with managerial intervention (Brosius 1999).

Nonetheless, indigenous peoples recognize themselves as "ecological" and practice an environmental politics in which both indigenous and non-indigenous peoples are welcome to participate and share in the effort to care for Mother Nature.

Chapter Seven

The Power of Ecological Identity: Alternative Ways of Thinking and Acting Within a Globalocality

INTRODUCTION

Indigenous peoples' environmental struggles and their construction of ecological identities have served as political strategies to establish bonds with transnational coalitions and networks (from financial help to political and conceptual support) that give indigenous peoples more political power within nation-states. These multiple, dynamic identities and loyalties (conservationist, NGOs, and indigenous nations, among others) locate indigenous peoples' movements within a new dimension of citizenship within the nation-state and as new agents in transnational eco-politics.

Indigenous peoples have found in western *ecological native* representations and environmental policies a means to gain recognition within national and transnational environmental discourses. They consider such representations opportunities to publicize their historical struggles to defend and recover their territories and as means to become autonomous and to maintain their traditional life plans. Despite all the many negative connotations and implications of *ecological native* representations, indigenous peoples' movements are using them to transform nonindigenous peoples' ideas of their identities not only within the nation-state, but also in transnational arenas. As we shall see, indigenous peoples' actions and identities and their interrelation with different nonindigenous actors have socioeconomic and political implications for a variety of actors and activities according to their specific social circumstances.

A multiform representational and discursive matrix addressing identity, culture, territory, social networks and political relations that transcends national borders constructs what I call a *transnational indigenous peoples' virtual ecological community.* This community forms a political arena that is changing the social reality in Colombia, promoting alternative ecological conceptions, and having national and international effects. Through the management of their own ecological identities, indigenous peoples have transformed western environmental disciplinary mechanisms (an eco-governmentality and its practices, discourses and representations) into mechanisms of indigenous resistance and action.

In this chapter, I focus on how the cultural and environmental politics of indigenous peoples affect different actors, including the state, multilateral institutions, transnational corporations, environmental nongovernmental organizations, social movements, local actors and researchers. Following Brysk (2000), I argue that internationalization and identity have been two of the most important means indigenous peoples have used to gain recognition and power within transnational politics.

In the SNSM, indigenous peoples' environmental politics and political actions have transformed their relationships with governmental and nongovernmental organizations in local and transnational settings. The previous chapters defined the international framework in which current relationships between indigenous peoples and the environmentalist have developed. Now it will be possible to specify how indigenous people of the SNSM have used their ecological identities within the local context to produce political change in local, national and international arenas. The power to produce these changes is particularly evident in Colombia's environmental policies and the programs and management plans of the FPSNSM and governmental institutions.

Indigenous peoples of the SNSM have positioned their cultural and environmental politics within regional political processes through the CTC. In this manner, they have also brought transnational agents into the local political process in an effort to integrate their strategies of modern science and planning: expert knowledge, sustainable management of natural resources, and the idea of territory in the modern sense of property. Indigenous policy proposals and political activities find a focal point in the idea of indigenous autonomy that has reshaped the terms of public discourse about nature, environment and ecology and formed the political context in which struggles over competing conceptions of nature, territory and property are taking place.

THE ROLE OF INDIGENOUS MOVEMENTS IN GLOBAL CIVIL SOCIETY

Local indigenous peoples' movements seek alliances with scholars, transnational indigenous peoples' movements and with ecological organizations, among others, in order to gain the power they need to claim "new rights" and build ties with environmental networks and academic communities in local and transnational settings. By doing so, local indigenous movements have created an ecological identity in national and international contexts.

Although these alliances with environmental organizations can cause contradictions and disagreements between indigenous and nonindigenous peoples, transnational environmental organizations have often provided crucial support to indigenous movements' efforts to gain recognition in international arenas as *ecological natives,* and this identity has proven to be one of the most effective strategies for the defense of indigenous peoples' territories and natural resources.

Indigenous peoples' transnational political links have contributed significantly to the formation of a global civil society that has impacted local civil society by expanding rights, mediating between local and state relationships, empowering local social movements and strengthening nonstate authority. The interests, ideas, conceptions, practices and identities that these political links mediate have, for example, made it possible for human rights organizations to publicize and denounce torture and discrimination at the local level to an international audience (Brysk 2000, Wapner 1995).

In the transnational civil sphere, indigenous peoples' movements are helping to form a stronger global civil society (Brysk 2000). Indigenous peoples' movements have become powerful agents that can transform social, political and economical processes at a global level precisely because they have established relations with other indigenous cultures and international regimes that transcend the nation-state (networks of NGOs, environmental movements and academic communities, among others).

In Latin America, the simultaneous emergence and confluence of democratic and indigenous social movements and nongovernmental and grassroots organizations has played an important role in the transformation of local, national and international political arenas and discourses. These movements, organizations and NGOs arose as responses to the authoritarian regimes and economic crises of Latin America. In some cases, they have replaced state administrative functions (Pardo 1997). In the Colombian context, there has been an unusual proliferation of NGOs. Pardo (1997) notes that in 1995 more than 70,000 NGOs were in existence. At this time, however, it is very difficult to know how many continue to function or have roles in Colombia because they appear and disappear according to their changing interests and levels of financial support. Nonetheless, they have provided indigenous peoples with technical support to effect the enlargement of democratic spaces within the state. These NGOs have thus become important parts of Colombia's civil society and state because they provide social movements the financial support to fulfill the functions that civil society or the state cannot fulfill.

Environmental NGOs have been powerful actors in the reconfiguration of "third world" ecological politics. There are various types of ENGOs (from first-world-based to third-world-based) with various environmental interests (from preservation to livelihood issues). ENGOs have become influential actors whose activities transcend national borders. Accordingly, ENGOs have gained influence; they have acquired the power to change corporate and governmental policies on a global scale and, therefore, the patterns of consumption of average citizens around the world. ENGOs have thus become a major link between local and the international political contexts (Princen 1994, Sethi 1993, and Wapner 1994, 1995).

However, ENGOs have had a mixed relationship with Colombia's indigenous peoples. Some have implemented sustainable development programs without consideration of local conceptions of nature or development. Therefore, it will be important to note the differences among ENGOs because a generalization of their function prevents us from seeing differences among them and their particular interests.

Other actors, such as scientific researchers (local or not) also influence decision-making about Colombia's environment. These actors form "epistemic communities" (Haas 1989, 1992) based on the distinctive discourses, practical techniques and disciplinary goals that inform their political actions in relation to environmental matters. Researchers in natural, social or technical sciences produce "expert knowledge" that gives them their power and status in political arenas. Thus, their recommendations and descriptions have had a great influence on the policies and activities of states, NGOs and grassroots organizations. In fact, some grassroots organizations have achieved their environmental goals because of the support they received from "expert scientific knowledges."

Social movements usually begin at the local level, but they often transcend the limits of their origins. Such transnational movements have been important for indigenous peoples' struggles since the 19th century. In fact, the first transnational association in support of indigenous peoples was formed in 1837 (Santos 1998). However, NGOs usually arise from national or international interest and only later do they seek out local opportunities to realize their objectives.[1]

In countries such as Ecuador, Brazil, Mexico, Nicaragua and Colombia, indigenous peoples and ecologists have led struggles against oil and timber companies, development programs (dams, highways) and bioprospecting. In Ecuador, the Huaorani people, with the help of transnational and national environmental movements, limited the DuPont-Conoco Oil Company's activities, and in Peru, similar coalitions helped indigenous peoples stop the Texas Crude Company. In Brazil (1989), the Kayapó people organized an international protest against large-scale development projects (Brysk 1993, 1994, 1996, 2000, Fisher 1994, Varese 1996, Collinson 1996, Conklin 1997).

In Colombia, during the 1990s, the Embera-Katio people protested the construction of the Urra hydroelectric dam on the Sinu River and the Uwa people were against the Occidental Petroleum Corporation exploration for oil in their territories with significant support from environmental movements that shared their concerns regarding the project's cultural and environmental effects. These are just two important examples of the relationships between environmental and indigenous movements, but they show how national and transnational ENGOs can provide crucial global support at the local level. These coalitions are also examples of the way that indigenous groups have used international support to make political changes at the national and local levels.

As for the Kogui people's situation, one of the factors that has contributed most to the transnational recognition of the importance of Kogui thinking and the SNSM has been Alan Ereira's film, "*From the Heart of the World: The Elder Brothers' Warning.*" In this film, he addresses the environmental bases of the Kogui thinking and their relation to the universe. Likewise, Ereira created the foundation "Tairona Heritage Trust" whose objectives are to spread the Kogui's messages and collect funds to recover Kogui ancestral territories in order to make them beneficiaries of the world's environmental concern.

Additionally, Ereira developed the "Tairona Heritage Studies Centre" attached to the University of Wales, Lampeter, in the United Kingdom, whose website has also spread the Kogui message and image throughout the world. In particular, it has disseminated an ecological representation of the Kogui as spiritual people very close to nature. In turn, this has generated other sites on the Internet that publicize the problems of indigenous peoples of the SNSM.

In this manner, the Kogui people travel through virtual space on the various sites that discuss their situation and way of thinking. They have thus obtained an outstanding position in global environmental iconography[2] that has generated international support[3] both material and spiritual, as in the case of the Ma'at internet website proposal that launched a call to stop using the names of indigenous peoples of the SNSM because revealing their names might make them targets of paramilitary and guerrilla forces.

The coalitions between indigenous peoples and NGOs also promote the international diffusion of indigenous peoples' claims for self-determination. At the same time, human rights, legal and peace commissions have played an important role in publicizing issues of indigenous peoples' rights. In Latin America, the Catholic Church has also been an important actor in forming and publicizing the goals of indigenous organizations.

According to Brysk (1993), these international regimes can make "a variety of contributions to social change". Also, she argues that these international regimes show that "social movements that lack conventional power can turn their weakness into strength by projecting cognitive and affective information to form international alliances" (261). In other words, she argues that coalitions between subnational and transnational actors can not only transform the state, but also create "information and legitimacy as a source of countervailing power in this process" (281).

PAN-ETHNIC TRANSNATIONALISM

Indigenous peoples have established relations with ethnic communities and indigenous nations around the world that transcend national boundaries in order to affirm a pan-indigenous identity (Bonfil Batalla 1981, Castells 1997, Brysk 2000). The coordination of these networks has been possible due to the transnational support for indigenous peoples who face similar environmental and cultural problems. These networks use the new communication technologies that allow rapid interaction with and access to support from indigenous peoples from different parts of the world. Most indigenous peoples' grassroots organizations now have webpages and links with other organizations that form a virtual indigenous network and community.[4]

There are many instances of the various and flexible relationships between environmentalist, indigenous social movements and ENGOs. In 1997, in the Amazonian region in Colombia, indigenous authorities (elders and shamans) from different parts of the world had a meeting to discuss their ecological strategies to protect their territories. An external environmental organization called Fundacion Sendama organized the meeting. In Caqueta, Amazonian healers decided to create the Union of *Yagé* Healers of the Colombian Amazon (*Unión de Médicos Indígenas Yageceros de la Amazonia Colombiana-UMIYAC*) to defend and preserve their medical practices, their territories and their forest. In June of 1999, when they celebrated the historic *Gathering of the Shamans* "Encuentro de Taitas en Yurayaco,"[5] they discussed the importance of the use of *Yage* to both their spiritual and medical practices. They also made a declaration related to the need to respect their territories and their collective intellectual property rights. Another result of this meeting was the Code of Ethics for the Indigenous Medicine of the Colombian Amazon.

In Santa Marta the first International meeting of indigenous people was carried out on September (27, 28 and 29), 1996, with participation from the Arhuaco, Wiwa, Kogui, Wayu, Guambiano, Uwa, Navajos, Quichuan and Shoshone peoples. One of the main themes was the sacred relationship that indigenous peoples maintain with their environment. Lorenzo Muelas, a Guambiano, noted that "indigenous cultures are just like a nonrenewable natural resource."

The "Second International Meeting of American Indigenous Peoples" that occurred in the SNSM-Nabusimake, in January 1999, had as its objectives the integration of indigenous peoples (Ijka, Maya, Kogui, among others) from different parts of the world and the promotion of their relationships with nature.[6] Currently, the third meeting is in a preparation process that focuses on traditional medicine.

The Kogui have established relations with the North American Natives as reported on networks (www.labyrinthina.com/kogi.htm, and www.tribalink.org), and these liaisons between peoples have produced reports and support concerning the indigenous peoples' problems, specifically environmental issues.

In the Latin American setting, we have to highlight the Yachac/Mamas, Ecuador-Colombia Meeting: Environment and Indigenous Worldview, held in October and November 1997 in Otavalo, Peguche and Ilum'n. There the Kogui peoples' participation was very significant in defining pan-ethnic strategies for indigenous peoples and their environments.

Examples of these indigenous transnational networks can be found on *Natiweb,* whose objectives are to protect the Mother Earth and to defend indigenous peoples' rights around the world, and on the *Indigenous Environmental Network*-IEN that promotes the protection of Mother Earth from contamination and pollution by strengthening, maintaining and respecting traditional techniques and "natural law." These are only two of many networks that articulate the relationships among different indigenous peoples and their relationships with the environment.

One of the most important indigenous peoples' organizations is the Tebtebba[7] Foundation, an Indigenous Peoples' International Center for Policy Research and Education. This organization was established in 1996 in Philippines. It "is firmly committed to the recognition, protection and promotion of indigenous peoples' rights worldwide" (www.tebtebba.org). Its legal work aims to influence "United Nations processes as they affect indigenous peoples' rights; monitoring the World Trade Organization, the multilateral financial institutions like the WB and the International Monetary Fund (IMF), the UN Convention on Biological Diversity (CBD) and other multilateral bodies" (www.tebtebba.org).

Tebtebba has established an international network of indigenous peoples through meetings and education programs that have found leaders and given them legal training. In Johannesburg, Tebtebba was important in placing the indigenous people's sentence in paragraph 22 bis. of the Johannesburg Political Declaration (September 4/2002) that states: "We reaffirm the vital role of indigenous peoples in sustainable development."

Victoria Tauli-Corpuz, indigenous activist from the Philippines and executive director of Tebtebba foundation, said, "We think the UN has taken a vital step towards respecting Indigenous Peoples equal to other peoples of the world." Indigenous peoples consider the inclusion of this term important in the legal context and as a historical moment.

These indigenous peoples' national and transnational networks that promote indigenous peoples' rights have intensified their efforts since the 1990s due to the technical and conceptual support of governmental and nongovernmental organizations and the efforts of the indigenous peoples' own organizations. In the Latin America context, COICA has played an important role in bringing indigenous peoples' demands to the international political arena.

INDIGENOUS MOVEMENTS' SELF-REPRESENTATION IN WORLD POLITICS

Indigenous peoples' demands for participation have been a constant in all the indigenous' declarations, although they achieved a higher level of coordination after the Earth Summit in Rio (1992). The articulations among indigenous peoples' demands and the United Nations' policies have created different international arenas for discussing indigenous peoples' problems. In this way, indigenous peoples have consolidated their participation in some political arenas (see table 7.1). However, one of the most important arenas for the globalization of indigenous movements was The First International Summit of Indigenous Peoples-B'okob' that took place in Guatemala in May 1993 and resulted in the Chimaltenango declaration.[8]

This declaration is very significant because it explicitly formulated the principal demands of indigenous peoples: to institute the Decade of Indigenous Peoples from 1994 to 2004; to reaffirm the rights of indigenous peoples to political, economic, social and cultural development based on self-de-

Table 7.1. Some Examples of International Arenas for Indigenous Peoples

SOME INTERNATIONAL ARENAS FOR INDIGENOUS PEOPLES (DATE OF BEGINNING)
United Nations-UN • Permanent Forum on Indigenous Issues (July 28/2000) • Working Group on the Draft Declaration on the Rights of Indigenous Peoples • Advisory Group of the UN Voluntary Fund for the International Decade of the World's Indigenous Peoples
Organization of America States-OAS • Working Group to prepare the Draft American Declaration on the Rights of Indigenous Peoples
World Summit on Sustainable Development-WSSD • Indigenous Peoples' International Summit on Sustainable Development
Convention on Biological Diversity-CDB • International Indigenous Forum on Biodiversity (1996) • The Ad Hoc Open-Ended Inter-Sessional Working Group on Article 8 (j) and Related Provisions of the Convention on Biological Diversity (2000) • Ad-Hoc Open-Ended Working Group on Access and Benefit-Sharing (2001)
Climate Convention and Kyoto Protocol • International Forum of Indigenous Peoples and Local Communities on Climate Change (2000)

termination and full participation in the decision-making process; to urge all governments to ratify all international instruments that promote respect for indigenous peoples' rights; to create the High Commission on Indigenous Peoples to ensure respect for indigenous peoples' rights; to create a Day of Indigenous Peoples of the World; to demand that the United Nations approve the Universal Declaration of the Rights of Indigenous Peoples and the respective ratifications and implementations on the part of various states; to demand that the United Nations continue and strengthen the Working Group on Indigenous Peoples as a permanent institution to monitor and ensure the fulfillment of the rights stated in the declaration; and to support a worldwide campaign against racism, among others.

In the Colombian context, there are also political arenas for discussing

Table 7.2. Some Examples of Colombian National Arenas for Indigenous Peoples

SOME COLOMBIAN NATIONAL ARENAS FOR INDIGENOUS PEOPLES
• National Environmental Council (Consejo Nacional Ambiental) • National Round Table for Agreement (Mesa de Concertación) • Regional Autonomous Corporations (Corporaciones Autónomas Regionales) • Specific Agreements between Governmental Institutions with each Indigenous Culture

the indigenous peoples' problems, for example, the National Environmental Council (Consejo Nacional Ambiental) and the Regional Autonomous Corporations (Corporaciones Autónomas Regionales) where indigenous peoples are active participants. There is also the national arena of negotiation (La Mesa de Concertación). However, in 2001, indigenous peoples abandoned this arena because they believed they could not use it to discuss in a deep way all their problems (see table 7.2).

In the case of the SNSM, the CTC's consolidation created an arena for discussing both the problems of the indigenous peoples and the SNSM's environmental situation. Within the national setting, participation from the Directive Committee of the PDS and the Regional Environmental Council has acknowledged the legitimacy of representatives of the CTC and promoted the strengthening of the indigenous government.

The CTC has also accomplished the establishment of a direct relationship with the WB to discuss the PAIDS' project and Global Environmental Facility's project to be implemented in the SNSM as well as relations with transnational NGO's such as the Nature Conservancy-TNC. However these projects are linked to activities of the FPSNSM. Therefore, all the negotiations are referred to agreements with the FPSNSM.

The CTC proposed that promoting and strengthening indigenous peoples' governments was the best way to conserve the SNSM'S environment. This proposal was approved unanimously by the Regional Environmental Council on December 10, 2001 and in the Broad Directive Council of the PDS on March 7 and 8, 2002, generating an agreement between the governmental institutions to respect and strengthen the CTC's policies. As a response to these agreements the Ministry of Environment issued Resolution No. 0621 dated July 9, 2002, which states:

First Article:
In the environmental planning and management process of the Ministry of the Environment, Special Administrative Unit of Natural National Parks, the Cesar Autonomous Corporation-Corpocesar, the Regional Autonomous Corporation of the Guajira-Corpoguajira and the Magdalena Autonomous Corporation-Corpomag, will incorporate themselves and develop priority themes of joint work agreed to by the Territorial Cabildo Council in the Extended Directive Committee of the Sustainable Development Plan of the Sierra Nevada de Santa Marta, held on March 7 and 8, 2002, as well as the ones agreed to in the future.

Second Article:
The Ministry of the Environment—Special Administrative Unit of Natural National Parks System and the Autonomous Corporation of the Cesar-Corpocesar, the Regional Autonomous Corporation of the Guajira-Corpoguajira and the Magdalena Autonomous Corporation, through its respective participation in the Sierra Nevada de Santa Marta, shall procure and promote strengthening of the Indigenous Government and the harmonic, integral and sustainable management of this strategic ecoregion incorporating traditional practices.

Likewise, the CTC's actions have caused an impact on the negotiating process with the FPSNSM, which publicly announced on August 28th, 2002 its acknowledgement of the "authority of the four indigenous peoples and their deliberation and decisions as represented by the Territorial Cabildo Council (Consejo territorial de cabildos-CTC)." Likewise, it proposes to defend and promote policies "so that the projects and actions set forth in the Sierra Nevada territory will have a measure of coherent participation, coordination and agreement with the CTC's policies and alignments."

In this manifesto, the FPSNSM subjects itself to Resolution No. 0621 dated July 9, 2002 from the Ministry of Environment, and it expresses its willingness to "construct a trusting and understanding basis with the Indigenous Organizations and their respective *Cabildos* that will allow them to guarantee full participation, and that their acts will positively contribute to and strengthen the indigenous government and the harmonic and sustainable management of the ecoregion of the SNSM".

Finally, the FPSNSM proposes "the construction of arenas for communication and coordination based on mutual respect regarding indigenous autonomy over the territory and that acknowledge indigenous territorial aspirations and everyone's responsibility to restore and conserve the SNSM."

The CTC, in response to the FSNSM's statement, on August 28, 2002, established the following coordination proposal as a condition of the FP-SNSM's intervention in indigenous peoples' territory in the Sierra Nevada. In this manner, the CTC's policies have accomplished a reconsideration of the arenas of multilateral coordination and the SNSM's conservation strategies. The CTC states:

> We have decided to provide an opening to dialogue between the CTC and the Foundation Pro Sierra Nevada, in order to agree on intervention policies for the Foundation Pro Sierra and its relationship with the people and indigenous organizations within the traditional territory. The historical events are treated as the CTC's main achievements contained in Resolution 0621 of 2002. In a decision approved unanimously by the Regional Environmental Council at its session held on December 10, 2002, there was a provision for "the strengthening of the Indigenous Government as an intervention strategy for the environmental conservation of the SNSM." In support of the Cabildo Territorial Council's policies, in agreements reached in the Directive Committee on March 7 and 8, 2002, among others.

CONCEPTUAL TRANSFORMATIONS RELATED TO DEFINITIONS OF INDIGENOUSNESS

In the last 70 years, in the international and national contexts, there has been a process of conceptual transformation regarding the definition of "indigenous" due to indigenous peoples' demands and their growing power in democratic political processes. In the Colombian case, these changes have allowed the Indigenous, Afro-Colombian, Creoles and Rom (Gypsy people) to be recognized as *peoples* (see tables. 7.3 and 7.4).

TRANSFORMING GLOBAL ENVIRONMENTAL DISCOURSES[9]

Indigenous peoples' conceptions of nature have influenced global environmental discourse by offering national and transnational ENGOs[10] conceptual tools to produce new relationships between society and the environment. In the same way, they have contributed to the redefinition of the strategies of

Table 7.3. Conceptual Transformations in the International Law Related to Definitions of Indigenous Peoples

CONVENTIONS, DECLARATIONS TREATIES	DEFINITIONS OF INDIGENOUS PEOPLES
2002. Johannesburg Declaration (Rio+10)	Indigenous Peoples
2002. World Conference against Racism, Racial Discrimination, Xenophobia and Related Intolerance	Indigenous Peoples
1995. Draft American Declaration on the Rights of Indigenous Peoples	Indigenous Peoples
1994. Draft Declaration on the Rights of Indigenous People of United Nations	Indigenous Peoples
1996. Decision 391 Common Regime on Access to Genetic Resources Andean Community	Indigenous, Afro-American and local communities
1992. Convention on Biological Diversity (Ley 165 de 1994)*	Indigenous and local communities
1989. ILO-169 (Ley 21 de 1991)*	Indigenous and tribal peoples
1963. Declaration on the Elimination of All Forms of Racial Discrimination (Ley 22 de 1981)*	Indigenous populations
1957. ILO Convention No. 107 (Ley 31 de 1967)*	Indigenous and tribal populations
1948. Universal Declaration of Human Rights (Article. 27).	Ethnic minorities

* Law by which that is ratified in Colombia

Table 7.4. Conceptual Transformations in the Colombian National Law Related to Definitions of Indigenous Peoples

COLOMBIAN NATIONAL LAW	DEFINITIONS OF INDIGENOUS PEOPLES
National Colombian Constitution (1991)	Ethnic groups and indigenous peoples
Law 21/1991 (Ley 21 1999)	Indigenous and tribal peoples
Law 89/1890 (Ley 89 1890)	Savages

natural conservation of those organizations. The emergence of environmental awareness, especially that of indigenous peoples, disrupts modern conceptions of nature and allows the creation of new or hybrid conceptions and meanings. This change has much to do with the articulation of environmental and indigenous peoples' discourses with developments in the natural and social sciences. These changes have helped to reconfigure the modern frontiers between nature and culture and to outline a new notion of nature as multiple and culturally constructed.

New notions that emphasize difference, plurality, fragmentation and complexity have transformed modern notions of linearity, universality, coherence and duality. These notions emphasize new thinking about organisms, reintegration with nature, active matter, uncertainty, chaos, probability, interpretation, and relativity of time and the space, among others. All these conceptual changes have affected "hard" natural sciences as well as social theories. Such new scientific perspectives on the universal and stable truths of modernity have created a dialogue among different sciences that has enabled the social sciences to formulate a critique of universal categories of progress, development, gender and nature.

In the course of these reconsiderations of the modern conceptions of nature, indigenous peoples have been important in bringing into the social sciences and modern thought in general new conceptions in which distinctions of nature and culture are inapplicable. Indigenous peoples' worldviews and conceptions have thus promoted visions of nature and humanity as interrelated.

Support for these views come from scholars in cultural studies, anthropology and ethnobiology, among others. These scholars affirm that humans have transformed the environment during millennia, carrying out selections on many species and influencing the regenerative cycles of the forests that affirm the interrelation and interdependence of all life. This is an ecological perspective in the whole sense of the term (the one that gives biology a way to express that species are related to one another as part of a unitary environment). In this way, the indigenous put in question a dualistic vision of nature as separate from a culturally-defined human species.

For some indigenous cultures, animals and plants have human behaviors and are regulated by social rules. In a reciprocal way, humans can become animals; these relationships between humans and nonhumans are in constant transformation and reciprocity. Under this perspective, the modern dichotomy of nature and culture becomes faulty. Indigenous peoples' conceptions do not correspond to western categories—although they have been in relation, interdependence and struggle with them—but to conceptions that articulate local knowledge as a combination of territory, identity and ethnic, social and cultural traditions. Such a view of humanity's place in life has greatly influenced environmentalist movements in general and, in particular, the biocentric tendencies within these movements.

The more immediate effects of the recognition of alternative visions of nature have to do with the possibility of generating management strategies according to alternative cultural practices. Consequently, western environmentalist strategies such as protected areas, national parks without people or scientific preservation, among others, have been questioned by alternatives that arise from indigenous peoples' local practices. In this way, the nonwestern systems of knowledge, especially indigenous knowledges, form an alternative response to the environmental crisis.

Redefinition of Strategies of Management of Natural Resources

Indigenous peoples have challenged western management strategies by proposing a new perspective on managing natural resources according to their own cultural practices which guarantee the alimentary and territorial security of indigenous peoples. In this way, they offer their knowledge as a solution to the problem of the conservation of natural resources in the world insofar as the greatest and rarest forms of biological diversity exist in their territories.

Beneria-Surkin (2000) describes how in Bolivia the High and Low Izozog organization (CABI) and their alliances with environmental movements have created wider political arenas for the indigenous peoples and the power they need to pursue decentralized strategies of sustainable developments for the Izoceño-Guaranies communities.

On a global scale, environmentalists and governmental programs have begun processes of redefinition of management strategies with local people's participation in an effort to protect biodiversity. These actions respond to diverse interests and political agendas, and so they promote a variety of activities: local participation so that the indigenous are agents of their own development and conservation programs; empowerment of local people so that they can negotiate with the local, national and transnational powers; and recognition of the rights, the knowledge and the culture of the local residents. However, indigenous peoples argue that the highest priorities for generating conservation strategies is the recognition of their self-determination and territorial rights, and the location of the discussion of the management of natural resources in a political arena rather than a purely environmental one (a managerial context in which the application of, for example, sustainable development or conservation plans might avoid political review).

In the Colombian case, for example, the superimposition of natural parks on indigenous territories has caused conflicts between indigenous peoples and the state. Indigenous peoples are opposed to the continual creation of protected areas in their territories. Consequently, the national government has begun a negotiation process that includes the proposals of indigenous peoples. Since 1998, the director of Natural National Parks Unit through its policy of Social Participation has worked in concert with diverse governmental and nongovernmental organizations, including indigenous peoples' authorities, in the formulation of different proposals for territorial and environmental planning in national parks.

The National Natural Parks Unit and ten indigenous grassroots organizations and twenty different indigenous cultures have generated agreements to construct cultural understanding related to environmental problems. In this way, indigenous peoples are opening spaces for the construction of management strategies that arise from their own conceptions and needs.

In the SNSM, indigenous peoples' discussions through the OGT and the CTC have confronted and had an effect on management strategies for the resources identified by the PDS. Their effect is evident in discussions that have occurred regarding how the environmental models of indigenous peoples could be implemented in the SNSM, not only through the practical continuity of indigenous forms of management, but also through the redefinition of practices in nonindigenous sectors such as rural agricultural.

The environmental practices of the indigenous people of the SNSM are very complex and reflect their cultural particularities (Kogui, Arhuaco, Wiwa and Kankuamo). However, in a general manner, there are certain philosophical principles, specifically promoted in the Kogui thinking, which support these practices and which are the ones that are assumed as general for the SNSM, not only by the indigenous people, but also by governmental organizations. These principles emphasize the need for indigenous autonomy and self-determination to enable indigenous peoples to manage their territories and resources in accordance with their cultural practices.

These changes in the conceptualization and the negotiations derived from the environmental and cultural politics of the CTC, and the ecological support for its conceptions and practices, have linked political self-determination to questions of territory, culture and identity. Accordingly, indigenous politics often focus upon the environmental practices that support the indigenous peoples' environmental management model in the SNSM.

The environmental practices of indigenous peoples of the SNSM cannot be considered in isolation from the cultural/spiritual plane of human/nonhuman relationships. As a matter of fact, the success of indigenous environmental practices depends on the successful observance of spiritual practices. The Kogui people maintain a spiritual identification with the "natural beings" that form part of their lives and so their relationships with them do not depend solely on practical environmental factors, a sense of ownership or material necessity. The territory is a vital entity and the parts that compose it, for example the mountains, are perceived as a live body. Each of nature's elements is a vital spiritual part of the indigenous culture that influences their relationship to their environment. The SNSM is the heart and center of the world, and through permanent care, the *Mamas* guarantee the universe's balance. Each landscape element (biotic and abiotic) has a symbolic and spiritual value, and so their territories contain many "sacred sites or places of origin" where the most significant events of mythical time occurred (FPSNSM 1997, CTC 2000).

Access to sacred sites is based on the Law of *Se* or The Original Law that governs the principle of harmony and equilibrium in the universe and that requires compensation for the use of nature. In this way, individual and collectivities generate environmental practices through the bounds established by the *Mamas* within the territory and through the spiritual entities in charge of the resources. These practices tend toward the equilibrium between material and spiritual planes, which consequently establishes a territorial management according to the region's ecological conditions. This management is based on knowing about the ecosystems and the biology of its species, as well as climatic and astral cycles among other factors, and all of which is framed as an interaction with spiritual entities. There is no separation of the natural and the spiritual worlds, but rather they support each other.

The *pagamentos* are material offerings that return as spiritual food to the Mothers the energy that they give to the human/natural world (Villegas

1999). It is the energy that is in permanent transformation and circulation through spiritual and material planes. When nature has a spiritual plane, it results in respect through ritual behaviors in relation to it that prevent disasters that may occur from not fulfilling these ritual behaviors. In this manner, environmental practices are not implemented with the purpose of conserving ecosystems or natural resources *per se,* but as social concepts that reproduce the indigenous human/nature scheme within an environmental/spiritual vision (Cayón 2002).

Environmental management and practices exercised by indigenous peoples in their territories are based on the following principles set out in the document "Alignments of Indigenous Peoples of the SNSM to Manage Their Territory" (OGT, UAESPNN and DGAI 2000):

- *Senunulang, niwi ka'gumu (Territory):* which was given to them as from the beginning, and is also a code that contains the standards that allows each individual to exist and that guarantees life's permanence. The *Mama* is the one that has the wisdom to understand it, interpret it and, from the knowledge of these standards, determine the use and management of the environmental relation.
- *Shibaks (The "Black Line"):* through the connection of sacred places, life's reproduction is possible. These places have to be interconnected to allow the care of nature, and through *pagamentos* spiritual communication and dialogue is permitted among the most important points in the territory. The Black Line is the territorial order that is kept by complying with the norms.
- *Ezwama:* the relationships among people and nature are practiced through it. As was already explained, it constitutes the collective principle of use and management in their own the territory. There are *ezwama* principles that concentrate the spiritual or material power that each lineage is responsible to care for. Each *ezwama* has a spiritual owner that, in turn, has a specific *Mama* who, according to the lineage (*tuke* o *tana*), is the one that has to take care of it."
- *Indigenous peoples' environmental authorities:* at the spiritual plane they have *Kalashé, Kalawia, Kalgwakwitchi and Kasougui. Ziníkula* is the spiritual father of the trees, from the period where everything was in *aluna.*[11] He is the one who gives an origin to the rest, to *Kalashé* and *Kalawia.* He is the main environmental authority, the owner that gives power to those *Mamas* in charge of organizing nature given by *Kalashé,* who is the one that organizes. He is the one who lets us learn about the science of environmental

> management. It is from there that we can ask for permission to cut down trees to build houses or hunt animals in the woods. The spiritual fathers such as *Kalashé, Kalawia,* and all the elders who own nature, must communicate in *aluna* with those other Fathers with whom they had to carry out works, such as *Kasougui.* These are persons that have to constantly communicate and be able to coordinate the different actions to be fulfilled. This coordination is effected in spirit, without any need to look at faces, so that even now many of the elder *Mama* can coordinate with each other in spirit, in *aluna.*

Approach to Environmental Practices from the Western Perspective

To discuss environmental practices of indigenous peoples of the SNSM from a western perspective is a very complex task: First, this is due to our very scarce and fragmented knowledge of them; secondly, it occurs because western knowledge views them from a perspective in which "different" phenomena are studied by different disciplines (biological and sociological, among others). In relation to the first aspect, indigenous peoples give their practices a sacred character and therefore keep them secret which prevents their disclosure. The few manifestations that have been observed are only the external manifestations of very complex relations. With respect to the second, the fragmentation of knowledge among western disciplines prevents us from being able to understand events in an integral manner.

Despite current interdisciplinary efforts to overcome these restrictions and appreciate human-nature interrelations, it is difficult to understand the complexity of a view that arises from cultures where such divisions do not exist and in which everything is integrated: the human, material and spiritual worlds. Table 7.5 identifies several of indigenous peoples' environmental practices in the SNSM. It is based on the bibliographic information collected by Cayón (2002) and from proposals made by the CTC to the PDS' Directive Committee (March 2002). Environmental practices shown separately are in fact complementary to each other and are involved not only in productive activities but also in a complex knowledges and systems of interrelation with other natural and spiritual elements. This chart at least suggests how these practices constitute a sustainable approach to natural resources and the conservation of SNSM's ecosystem, even though it has the limitation of a western point of view.[12] As the chart indicates, indigenous peoples' environmental practices also take form as spiritual activities which reflect their knowledge about ecological interactions, species biology, biotic components (climate, soil), and astrology, among other aspects. This knowl-

Table 7.5. Some Indigenous Peoples' Environmental Practices in the SNSM

	ENVIRONMENTAL PRACTICES	WHAT THEY CONSIST OF
RELATES TO SOCIAL AND RITUAL DYNAMICS	Lineage distribution and basins management	Taking care of basins totally
	Establishment of sacred places	In several places this coincides with key ecological sites, as in the lakes of the high plateaus. These act as reservations or protected areas
	Ritual restrictions, a set aside of several species to feed *Mamas*	Reduces pressure on species
	Population mobility by spiritual activities	Distributes pressure on different areas and ecosystems
	Resource-use control through the *Mamas*. Related to use and consumption restrictions in certain circumstances	Controls pressure on species and other resources
RELATED TO TERRITORIAL MANAGEMENT	Use of a different thermal grounds (between 300 and 3500 m.); and use of various ecosystems during the year; mobility through different territories and thermal grounds	Distributes pressure upon resources and soils in specific places. Allows the diversified use of resources and a variety of products. Integral management of different types of ecosystems
	Polyactive production	Diversifies subsistence activities and thus distributes pressure on resources

(*continued*)

Table 7.5. (*continued*)

	ENVIRONMENTAL PRACTICES	WHAT THEY CONSIST OF
	Soil and slope management	Good use of rain without soil detriment
	Agricultural cycles	Allows soils to rest and renew their nutrients; controls erosion and drainage. Allows natural regeneration
	Cultivated and non-cultivated zone mosaic	Management of game animals allows more species access to gardens and maintains the local fauna
	Polycultivation	Protects the soil from erosion and allows select plants to grow
	Natural reforestation	Natural forest recovery
	Use of soils according to their vocation	Allows the adequate use of each type of soil
	Vertical use of different thermal floors	Allows the distribution of pressure over resources of different Orobiome and the reporting of a range of use and consumption products
RELATED TO RESOURCES MANAGEMENT	Zones closed to hunting	Reduces pressure on animal populations
	Broad use of animal species	Distributes pressure on species

(*continued*)

Table 7.5. (*continued*)

	ENVIRONMENTAL PRACTICES	WHAT THEY CONSIST OF
	Rotation of hunting areas	Distributes and alternates pressure within different animal groups

edge allows indigenous peoples to efficiently manage natural resources according to ecological conditions in order to distribute pressure among resources, habitats and geographical areas.

The indigenous peoples of the SNSM also consider the following mode of territorial ordering essential to the function of their culture as part of nature: The SNSM as a territory delimited by the Black Line that extends from the lowlands to the highlands and around the river basins as wholes. This is so because the interrelation of human activities at the different altitudes of the SNSM with natural elements and events constitutes a spiritual unity.

Consequently, human environmental management has a good reason to be considered a natural event according to the indigenous standards for human/nature actions (*pagamentos,* confession, *sewa's* exercise, use and protection of *ezwama* lands, and support of *shibaks* sites), all of which serve to assure that human beings maintain the future viability of the environment. These standards acquire political authority in the *Mamas* who make clear the responsibility that every person shares for the continuance and well-being of the universe.

This view of natural and cultural interrelation has permeated the governmental decision-making authorities (the National Planning Department and its environmental management directorship, the Ministry of Environment, and Regional Autonomous Corporations) that formulate environmental policies for the SNSM, thereby creating an arena of conceptual discussion that rethinks the western categories of nature and human interrelations with it. At the same time, this repositions indigenous peoples of the SNSM in the local context as authoritative proponents for alternative forms of the environmental management.

COUNTER-GLOBALIZATIONS

Santos (1998) argues that indigenous peoples' demands for collective rights of self-determination allow alternative forms of law and justice and new modes of citizenship that could be considered parts of a counter-hegemonic globalization. In fact, indigenous peoples' collective rights to self-determination go beyond the state toward local-transnational bonds. Their view of rights questions the notion of sovereignty of the state by demanding a new kind of sovereignty based on a sharing and polyphonic multiculturalism that transcends nationalism. Finally, indigenous peoples' self-determination implies a communitarian life, but one linked to transnational processes.

Santos (1998) argues that globalization "consists of sets of social relations" and "as these sets of social relations change, so does globalization" (56). Moreover, he notes that there is, strictly speaking, no single entity called globalization; there are, rather, globalizations. Thus, he proposes as a definition of globalization "the process by which a given local condition or entity succeeds in extending its reach over the globe and, by doing so, develops the capacity to designate a rival social condition or entity as local" (Santos 1998:56). In fact, he distinguishes four modes of production of globalization: globalized localism, localized globalism, cosmopolitanism and the common heritage of humankind. Globalized localism consists "in the process by which a given local phenomenon is successfully globalized" (Santos 1998:57); and localized globalism involves "the specific impact of transnational practices and imperatives on local conditions that are thereby destructured and restructured in order to respond to transnational imperatives" (Santos 1998:57).

However, Santos (1998) argues that the intensification of global interactions also entails two other processes: cosmopolitanism and the common heritage of humankind. Cosmopolitan activities imply transnational relations and solidarities among individuals and organizations (from academic to human rights groups) that promote subversive counter-hegemonic globalizations. Finally, the common heritage of humankind implies global actions related to, for example, environmental changes, preservation of biodiversity and conservation. The international community administers these global actions on behalf of future generations in a manner that transcends nations and hegemonic globalisms to propose a variety of alternative discourses and practical options. However, as noted before, this conception of global heritage is not necessarily beneficial or liberating to indigenous peoples.

For Santos (1998) cosmopolitanism and the common heritage of humankind are globalizations from below or counter-hegemonic globalizations. These are expressions of resistances from, for example, grassroots organizations, indigenous movements and democratic political organizations that "try to counteract social exclusion, opening up arenas for democratic participation, for community building, for alternatives to dominant forms of development and knowledge, in sum, for social inclusion" (1998:58). These two globalizations disrupt hegemonic globalizations (primarily neoliberalism) by promoting participatory democracy, alternative production systems, and emancipatory multiculturalism, justices and citizenships that can strengthen civil society.

In a similar way, Leff (2002) states that it is the market-oriented logic of neoliberal economic values that has generated worldwide socioenvironmental degradation. In response, there is "emerging a politics of the place, space and time (Leff, 2001c) mobilized by the peoples' new rights to cultural identity [. . .], legitimizing more plural and democratic rules of social interaction." Under this perspective, indigenous peoples and their interrelation with their territories have been promoting processes of interrelation with and control of nature that give rise to options that differ from those of economic globalization.

> A new politics of place and difference has been built from the sense of the time in the current struggles for identity, autonomy and territory. What underlies the clamor for the recognition of survival, cultural diversity and quality of life's rights of the peoples, is the politics of being; it is a politics of becoming and transforming that valorizes the meaning of. . . each individual and community right to forge their own future. (Leff 2002:13)

FINAL REMARKS: RETHINKING MODERNITY?

Contemporary discussions about Latin America often characterize it as participating in non-modernity, premodernity, uneven modernity, or modernization without modernity or postmodernity. Similar debates have also framed the discussion in Colombia. I cluster these discussions under four main approaches: (1) Conventional analyses that consider Latin-American failures to become modern a result of Latin-American backwardness; (2) Analyses that propose that international political and economic powers have helped to produce a particular Latin-America modernity that does not follow the classical or typically western paths to modernity; (3) Postmodern analyses that explain Latin-American sociopolitical situations as diverse and multiple in ways that break with or differ from the linear and simultaneous progressivism of the modern process; and (4) Alternative analyses particular to Latin America sociocultural formations.

Theoretical concerns about the adequacy of concepts such as modernity, postmodernity and premodernity, have led some to put forth alternative theoretical visions to enable the analysis of the local cultural practices and particular historical processes that characterize the Latin-American region. Some scholars have constructed new paradigms that give a new understanding of Latin-American identities (distinct from what is described as "postmodern") that arise from the particular realities in Latin America. For example, Latin America has begun to be analyzed in terms of its hybrid cultures (Garcia Canclini 1995), metamorphic identities (Bartra 1992), disjunctive democracies (Caldeira 2001), and alternative modernities (Escobar and Pedrosa 1995). All these approaches share in common the effort to analyze the processes of construction of Latin-American subjectivities and identities in relation to both national and transnational processes. Moreover, these approaches seek to understand sociocultural diversity—within Latin America and within each country—and provide particular proposals to deal not only with the modern project but also with Latin America's specific historical processes. Among such scholars are some who propose to rethink modernity and Latin America at conceptual, practical and technical levels.

For example, Garcia Canclini (1995) proposes that pre-Columbian, colonial and modern traditions have interacted to produce successive hybridizations that do not constitute one particular social system or subjectivity but rather different social systems and particular subjectivities. In this way, Latin Americans can travel through not only different spaces and cultural practices (traditional/modern), but also can travel through time/space relations that constitute "procedures of passage from one situation to another." Therefore, he claims to see those fluid processes as an integral part of Latin American identities rather than conflicting elements in a contest between the traditional/modern and the modern/postmodern.

In a similar way, Bartra's (1992), analysis of Mexico indicates that Mexican national identity is as a result of hegemonic discourses that suppressed social dynamics whose diversity challenged the ideal of a unified nationality and nation. Thus, he uses the idea of metamorphosis (through the metaphor of the axolotl) to rethink the idea of a total Mexican unity, and to propose new readings of Mexican realities based on the heterogeneity and change, often suppressed by hegemonic nationalist discourses.

Caldeira (2001) argues that analyses of citizenship require understanding it as a part of modern thought that has passed through long processes of transformations from the uncontrolled "primitive" body to the acquisition of individual rights as a subject within the nation-state: "The history of the passage from the dominance of the canon of the grotesque body to that of individual body in Europe is crucial for the formation of modernity: it signifies the hegemony of new sensibilities and cultural values, the triumph of new forms of social relations and social organization, and the establishment of new forms of control and subjection" (Caldeira 2001).

Therefore, the modern ideal of citizenship has not been completely accomplished in Latin America because the implementation of social membership through citizenship expresses a particular historical process which took place in Europe but that could not be reproduced in the same way in Latin America. Moreover, specific sociohistorical conditions in Latin America allowed a different interpretation and appropriation of these "universal" ideals.

Caldeira (2001) also stresses how the developmental progression of this historical process (European modernity) has been used as a model to analyze Latin America. Therefore, scholars who have made analyses of citizenship under this general idea look for sequential society-wide advances that cannot be found in Latin America. "Countries such as Brazil, but also others with different histories (usually colonial histories) and that have today disjunctive democracies, force us to dissociate the elements of that history and to question its model sequence" (476).

Alvarez, Dagnino and Escobar (1998) argue that the cultural politics of the social movements have located their proposals within the nationalist projects of Latin America. In this way, these social movements have gained the political power to reshape and to rethink predominant concepts of nation, citizenship, development and democracy that constitute modernity itself. As Albo (1995) says, "[t]he proposals of Indian organizations are undoubtedly a project for the reformulation of society itself, no matter how embryonic their present formulations are. They have found acceptance, as well as hesitation, in the upper strata of our societies" (24). In this way, indigenous peoples are promoting a pluralization within the ostensible unity of the nation-state.

Escobar (1992) relates the emergence of social movements to the crisis of development as a cultural and economic project of modernity that became evident in Latin America. He notes how social movements, especially grassroots movements, have had to respond to this crisis by developing the potential to postulate alternative modes and conceptions of development and, by doing so, create changes in the cultural and economic assumptions of modernity itself. Escobar observes how the crisis of modernity has produced new social actors and mechanisms for "the production of meanings, identities, and social relations. Social movements must be seen equally and inseparably as struggles over meaning as well as material conditions; that is, as cultural struggles" (Escobar 1992:69). In a similar way, discussions of political ecology qualify social movements' environmental politics as challenges to modern notions of development. Therefore, indigenous peoples' movements, especially their environmental components, are challenging modern notions of citizenship, democracy, development and nature.

The dynamics of the cultural politics of indigenous movements have helped to resignify notions of nature and development and, by extension, modernity in the West by opening new democratic arenas and changing the ways of doing politics. In Colombia, this new political context in which indigenous peoples have achieved a degree of autonomy has opened a space for constructing *indigenous ecological collective identities*. These collective ecological identities have found support in national and international contexts through recognition of the high biodiversity that result from indigenous cultural practices and methods for managing natural resources.

In contrast, "developed" areas have been transformed into relatively sterile lands because of massive economic production. The crisis of development and the increasing depletion of natural resources evident in industrialized societies have resulted in a redefinition of "the environment" (from a source of production of raw material to space of reproduction of life itself) that has given indigenous peoples the authority to rethink modern notions of nature, development and industrial processes.

Some scholars (Escobar and Pedrosa 1996, Grueso, Rosero and Escobar 1998) argue, for example, that social movements in the Pacific region in Colombia based on collective identities (Afro-Colombian and indigenous peoples) have promoted new conceptions of development that challenge the idea that western development programs are the best means to achieve industrialization, mechanization, or improved standards of living.

These movements and their proposals challenge neoliberal notions of development by rethinking "nature" and proposing new strategies for managing ecosystems. In fact, indigenous peoples have made new models of development for themselves called "life plans." In these plans, indigenous peoples propose maintain and disseminate their own way of thinking and understanding the relation between humans and nonhumans. In these plans they discuss their political and cultural interests, different strategies for the management of natural resources and their territories, as well as new perspectives and alternative forms of development that consider more than economic growth and profit.

In general, some indigenous peoples, ethnic minorities, social-movement members, scholars and technicians now advocate the view that there cannot be a fixed and rigid order that predicts the processes and progression of "civilization." On the contrary, they consider that the dynamics and contingencies of Latin America create opportunities to explore alternatives to the standardized and problematic progressive conceptions of modernity, modernization or postmodernity derived from the West. However, as was we have seen in previous chapters, even multiculturalism is a part of a western-dominated governmentality, and western environmental concerns have given rise to a globalization of politics that introduces indigenous peoples to an eco-governmentality whose values and proposals are often at odds with those of indigenous peoples.

Chapter Eight
Indigenous Peoples Within Eco-Governmentality

INTRODUCTION

Instead of offering conclusions, this chapter presents some final thoughts about the contradictions, paradoxes, and implications of the relationships between indigenous peoples and environmentalism. These relationships are in constant transformation, and it is difficult to say if they have produced any clear victories or losses for indigenous peoples. Nonetheless, I want to highlight some specific outcomes and consequences concerning the introduction of indigenous peoples into new eco-political and eco-economic processes. The fact that these outcomes and consequences so often result in contradictions between indigenous and nonindigenous values and complicate the achievement of the goals of all involved certainly puts in question any triumphalist analysis of indigenous peoples' movements. Of course, these processes are ongoing and open-ended, and so I can produce no final answers here regarding their outcomes. What is clear, however, is that Indigenous peoples can use and transform eco-governmentality to serve their own ends and possibly those of nature and humanity, although they face many difficulties in their efforts to do so.

Indigenous peoples' movements can be seen as victorious in having positioned their ecological identities in national and transnational environmental discourses and global eco-politics. They have indeed succeeded in establishing bonds, coalitions and networks (with local, national and transnational governmental institutions, environmentalists, NGOs, human-rights organizations and other indigenous organizations, among others) that give indigenous people a means to represent their interests more effectively and gain more political power within nation-states (chapters 2, 3, 5 and 7), especially with respect to their primary goals: to assert sovereignty over their territories and assert political autonomy and self-determination with respect to defining the values and goals of their life plans. In this sense, it may be argued that indigenous peoples have indeed been victorious. However, a closer examination of the results of indigenous peoples' efforts to promote their goals and gain recognition in national and transnational arenas reveals that those efforts have often resulted in mixed or contradictory outcomes. The reasons for this can be found through a close examination of how *ecological native* representations and the values attached to them have been used by both nonindigenous and indigenous interests to vindicate often diametrically opposed definitions of political, economic and environmental problems and approaches to their solution. From this perspective, it is not at all certain that indigenous peoples have succeeded in perhaps their most ambitious goal: to transform the West's modern and neoliberal approaches to indigenous territorial and cultural goals and global environmental problems and solutions (chapters 4, 6 and 7).

The processes of economic globalization that affect biodiversity mean that the relationship between indigenous peoples and environmentalism has to be analyzed according to new relationships of power/knowledge due to the globalization of the environmental strategies designed to "protect" nature and promote eco-governmentality (chapters 4 and 6). However, within the new western version of the environmental regime, or eco-governmentality, neoliberal values are often reaffirmed to the extent that eco-governmentality encourages indigenous peoples to transform their relationships to their knowledges and the resources in their territories according to economic conceptions of commodity and profit that are in many instances completely alien to their values and goals. Even western goals apparently compatible with those of indigenous peoples, such as environmental conservation or preservation (global heritage), prove to be problematic from the indigenous perspective insofar as they are infused with modern and neo-liberal assumptions.

The economic aspect of these issues can be seen in the following example:

> Four Vancouver Island First Nations have been awarded over half a million cubic metres of timber—enough to build 25,000 houses [. . .] The award marks the first major resource allocation under a new law aimed at increasing aboriginal participation in the forest economy. [. . .] The award also signals the government is willing to use the undercut as part of treaty settlement processes. The four First Nations have signed interim measures agreements

with the province in which Victoria committed to provide them with timber. (Gordon Hamilton, *Vancouver Sun*, Wednesday, January 29, 2003).

As is evident above, indigenous peoples' ecological identities and interests can be constructed and represented in manners that promote neoliberal conceptions of territory, resources and environmental rights: property, commodity, and contract. Such western ways of perceiving the *ecological native* and their territories create new relationships for indigenous peoples that link them to western economic processes that have generated new conceptualizations of nature as a human heritage and as an eco-commodity. These relationships can create tangible economic benefits for them, but they often do so in a manner that undermines indigenous peoples' efforts to pursue their more encompassing and properly indigenous political and cultural goals, particularly those involving sovereignty and self-determination within their territories (table 8.1).

Table 8.1. Contradictions and Effects of Eco-governmentality

TOPIC	CONTRADICTIONS	EFFECTS FOR INDIGENOUS PEOPLES
❖ Conceptions of Territory and Sovereignty	⋎ National territory	→ The state is sovereign over its territory and resources, so its has to deny indigenous peoples' rights previously recognized
	⋏ Indigenous peoples' territories	→ Indigenous peoples' rights of self-determination will allow a direct negotiation with transnational corporations
	⋎ National property and intellectual property rights	→ Genetic resources are national, so indigenous peoples cannot negotiate their resources
	⋏ Indigenous peoples' collective property rights	→ Indigenous peoples have collective property rights so they can sell their resources without the state's control

(*continued*)

Table 8.1. (*continued*)

TOPIC	CONTRADICTIONS	EFFECTS FOR INDIGENOUS PEOPLES
❖ Conceptions of Nature	⋎ Cultural construct	→ Indigenous peoples are the owners of their resources, so they can decide what to do with them
	⋏ Natural thing	→ Indigenous peoples are not the owners of their natural resources, which are "natural" so that they are a global heritage
	⋎ Preservation of nature	→ Indigenous peoples' territories are biological empty spaces
	⋏ Selling nature	→ Indigenous people's territories are introduced within eco-economic circuits
	⋎ Ecological knowledges and traditions to be recovered	→ Indigenous peoples' knowledges related to environmental practices are reconstructed and reinvented
	⋏ Ecological knowledges and traditions of the other	→ Ideal images of noble savage are imposed
❖ Separation of Indigenous Peoples and Environmental Rights	Indigenous peoples' rights: ⋎ They are not implemented by all the states, and they are not recognized in all their extension	→ Indigenous peoples are not fully recognized as *peoples*
	Environmental rights: ⋏ Environmental rights are based on modern notions of nature	→ Indigenous peoples' territories are affected because their resources are now considered a global heritage
	⋏ Environmental rights are not recognized	→ Environmental strategies are implemented to ameliorate development impacts

DILEMMAS OF ECO-GOVERNMENTALITY

There are some important points that I want to address regarding the actual social, economic, political and cultural situations that confront indigenous peoples, in general, and indigenous peoples of the SNSM, in particular. The emergence of an eco-governmentality related to global environmental policies (chapter 4), and the recognition of multiculturalism (chapter 2 and 3) imply new situations and contradictions for these peoples of which three are pertinent in the present context: the relationship between national sovereignty and indigenous self-determination; the emergence of new conceptions of nature; and the separation between indigenous peoples' rights and environmental rights (see table 8.1).

National Sovereignty versus Indigenous Peoples' Self-determination

One of the principal contradictions between indigenous and western peoples is that the new international environmental policies reinforce the idea of individual property rights in natural resources while also recognizing the collective autonomy of indigenous peoples with regard to their territories. The recognition of biodiversity as a new commodity that can be measured, priced and bought creates new political, economic and cultural issues for indigenous people. These situations have given indigenous people a powerful presence in international arenas because they have territories with valuable resources such as biodiversity. Consequently, nonindigenous economic agents want indigenous peoples to have rights over their resources and political sovereignty over their territories so that they can make binding agreements to sell them to the highest bidder.

The Search for New Raw Materials and Genetic Resources

Biodiversity, however, is geographically located mainly in "third-world" countries, therefore "third-world" countries claim their national sovereignty over their territories and natural resources in order to force transnational corporations to negotiate with the nation-state rather than indigenous peoples. In addition, "third-world" nation-states assert their sovereignty over "national" territory to resist neoliberal transnational policies. Even though nation-states often implement policies that transcend the model of nation-state unity and sovereignty when they permit decentralization and transnational participation at the local levels, as in the case of indigenous peoples, they also have to resist those policies, otherwise they lose their sovereignty over biodiversity.

This interplay between the international and national policies demonstrates how indigenous peoples' rights and a global environmental law based on property rights designed to protect biodiversity can come into contradiction with the idea of national sovereignty over natural resources. National laws in fact recognize the idea of multiculturalism and indigenous peoples' rights to decide the management of natural resources in their territories, which means that those peoples can deal directly with transnational corporations and erase the state's power in those negotiations. At the same time, indigenous peoples have been empowered to negotiate with transnational corporations regarding their natural resources as nation-states, which contradicts neoliberal policies of the reduction of states' power by creating (recognizing) multiple micro nation-states with negotiating power at a micro scale.

In the case of Colombia, the national law is contradictory because it allows indigenous autonomy despite the fact the national state does not want to lose its sovereignty over "national" territory, especially territory with genetic resources or tourism potential. In the case of the Embera-katio people, the state responded to the petition of indigenous peoples, in the following manner:

> To recognize [indigenous peoples] as Environmental Authorities, especially the biggest indigenous peoples' councils, it is necessary that the conditions and functions of these environmental authorities are clearly defined in the laws and the same ones should determine any modification; in consequence this is a serious matter of competition with the Congress of the Republic. In spite of the above-mentioned, it should not get lost from view that in special situations, for example when areas of the system of national parks are shared areas of indigenous peoples' territories, the community carries out some functions of control of environmental authority that are suited to its own authority in fact.

The state thus continues to refuse indigenous peoples' rights of autonomy and territorial control; for if the state allowed self-determination in the territories, it would imply the loss of state sovereignty over natural resources and the environment. As Escobar points out (1999), the indigenous communities are being recognized as owners of their territories, but only to the extent that they treat their territories like "environmental capital" useful for national development purposes.

Criticism of the nation state's resistance to indigenous conceptions of territorial rights appears in the following statement from the Working Group on the Draft Declaration on the Rights of Indigenous Peoples:

> Many indigenous representatives stressed the fundamental importance of self-determination in the draft declaration. Grave concern was expressed that a significant number of States in the Working Group are currently failing to adhere to the Purposes and Principles of the U.N. Charter. Five of these States constitute as well the five permanent Members of the U.N. Security Council and should be role models in terms of upholding the basic values and principles of the U.N. Charter. Indigenous representatives stated that self-determination is an accepted human right and that the addition of the term "internal self-determination" as proposed by the United States or other proposals that might restrict the right had no basis in international law and were discriminatory. In regard to natural resources and indigenous peoples, the United States position was also said by many indigenous representatives to be discriminatory and inconsistent with the specific conclusions of the U.N. Human Rights Committee and with international law. In particular, Indigenous representatives strongly rejected any interpretation of article 1, Para. 2 of the international human rights Covenants that suggests that economic aspects of the right of self-determination—namely, the right to natural wealth and resources—would simply not apply to indigenous peoples. It was added that the denial by States of the resource rights of Indigenous peoples is a major reason for the enduring legacy of severe impoverishment inflicted on indigenous peoples globally. (The American Indian Law Alliance (AILA) reviewed the unedited final draft of the report of the eighth session of the Working Group on the Draft Declaration on the Rights of Indigenous Peoples (E/CN.4/2002/92) as posted on the website of the UN High Commissioner for Human Rights, December 30/2002)

These contradictions are also evident among the indigenous peoples of the SNSM, who have been unable to consolidate their political autonomy and self-determination over their territories and resources because of the overlapping national and global policies that affect their territories and resources and introduce the SNSM to international circuits of eco-commodities. The nation-state thus demands sovereignty over natural resources, despite the fact that it has recognized indigenous peoples' autonomy in the national constitution.

Although indigenous peoples in Colombia have the right to control their resources directly without state mediation, some environmentalists argue that it would be easier to enforce national and international environmental policies under state control. In contrast, other environmentalists argue that local and national governments are corrupt, so there is a need to introduce global environmental laws and legal institutions into these territories through a global eco-governmentality. Some even claim that it is better to have transnational corporations directing indigenous and national environmental policies and practices under the concept of enlightened self-interest.

In Colombia, the national government, through constitutional mandate and other laws, such as 99 of 1993, has allocated economic resources to generate participation and agreement processes that identify and resolve the environmental problems of various social groups. Thus, both the state and indigenous peoples have jurisdiction over their territories and decision making in relation to them. This dual jurisdiction is recognized politically and legally. However, indigenous peoples, in addition, have ancestral rights from their own laws and cultural practices. These conflicting jurisdictions affect the autonomy of indigenous peoples in the management of their territory.

In addition, the state's policies have begun to back away from granting indigenous peoples' rights. The offices of indigenous affairs have been reduced around the country. The policy priorities of the state now focus on recovering sovereignty, fighting paramilitaries and guerrillas, following the recommendations of the International Monetary Fund and fulfilling the World Bank's guidelines for the control of illegal crops.

The recognition of indigenous peoples as *ecological natives* (although it generates many contradictions as we have seen in chapter 6) is not necessarily an impediment to national and transnational economic powers in their encounters with indigenous peoples who assert rights of self-determination and autonomy in their territories. Since indigenous territories have a great biodiversity, as well as minerals and petroleum, there has been a national and international intervention in indigenous territories (as in the cases of the OXY in Uwa's territory and the construction of the Urrá hydroelectric dam in Embera-katio's territory). Despite indigenous objections and the national policies addressing indigenous rights and environmental protection, big development projects continue being planned for indigenous territories without previous consultation with indigenous peoples.

The new economic valuation processes involving biological diversity have generated much interest regarding indigenous peoples' territories. Consequently, there is a search for new raw materials and genetic resources throughout their lands. Their territories have unprospected oil and mineral reserves because the nation has long ignored their economic exploration and development due to their previous unimportance to the central state and their distance from the national urban centers. These factors left these regions at the margins of modernization. Now, these same factors have repositioned these territories as focal points of national and global interest as new and important sources of raw material and genetic resources.

Indigenous peoples seem to have autonomy and rights over their resources, which means that access to genetic resources and their ownership have to be established under the new legal and institutional parameters of biodiversity agreements that require indigenous participation. However, indigenous control and access to these resources cannot be direct because the state also asserts national sovereignty over these genetic resources and because a unique transnational regulation model has been developed within the international arena that includes the ILO-169, the CDB Agreement (8j and 10c) and the Andean Decision 391 of 1996.

These conflicting systems of governance affect all aspects of indigenous political life in Colombia: territoriality, governability, autonomy, self-determination and law. At the same time, they affect the indigenous peoples' cultural relationship to "nature," which has multiple representations and dimensions that are not necessarily related to western economic, biological or genetic considerations.

In the international context, the following schemes have been developed to "resolve" this problem: the generation of general guidelines for direct contracts between transnational corporations and the indigenous community that has genetic resources; the creation of *sui generis* regulations; and the reduction of discussion regarding sacred sites. In these scenarios, the common idea is that scientific and commercial access to genetic resources is necessary and desirable and should be expedited regardless of indigenous objections. Indigenous knowledges are thus still viewed under the light of commercially-influenced scientific knowledge (ICANH-Institute von Humboldt-Colciencias, 2002).

Indigenous peoples' positions regarding the resolution of the issue of access to genetic resources have been quite different: a categorical NO to access if their knowledges and intellectual and territorial rights are not taken into account; and a provisional NO until the creation of ethical codes or *sui generis* legal systems of protection that do take into account their values and interests.

For example, in Peru and Bolivia, a previous consultation of indigenous peoples resulted in the following proposals for access to genetic resources: 1) Internal sovereignty of indigenous peoples and local communities with subsequent internal negotiation, generation of community records and creation of *sui generis* legal systems; or 2) national control of the access to genetic resources negotiated with the state following the CAN model (main contract, accessory and attached agreements) to produce binding national documents that employ national intellectual property rights' guidelines.

In Colombia, current discussions of this issue revolve around the CBD (8j and 10c) and the Andean Decision 391, specifically Articles 7, 8 temporary and 9 temporary. However, Decision 391 has problems since it has not been implemented or regulated. It has gaps and contradictions, and needs a political action framework that provides guarantees of indigenous peoples' rights, culture, territory, and knowledges. Also, there has not been any previous consultation or dissemination of the issue within indigenous peoples' communities. Moreover, the legal context of the CPC-91 and the ILO-169 has not been recognized, and there are no national consultations that involve ethnic groups and local communities in the discussion of this issue. Therefore there is a conflict of interests between the national state and indigenous peoples.

Access to genetic resources is a theme that has not been sufficiently discussed within the context of indigenous peoples' policies in the SNSM. However, that topic is now the order of the day since it is related to biological research processes that are ready for implementation in the SNSM through the GEF program. Indigenous peoples have therefore decided, as a protective mechanism, to refuse any biological research process in their territories for the foreseeable future.

In addition, the SNSM's indigenous peoples' territorial autonomy, as well as that of other indigenous peoples, has been complicated by the activities of various local, national and transnational agents. These activities involve new relationships with land and resource management that contradict indigenous peoples' environmental beliefs, such as the production of illegal crops of marijuana and coca and their consequent effects: violence and displacement. Indigenous peoples are also facing the loss of sacred places in the lowlands that breaks the communication with the highlands, which is fundamental to *pagamentos* and the spiritual dialogue among the territory's sacred sites (OGT-UAESPNN & DGAI 2000). All these issues affect the autonomy and self-determination of indigenous peoples.

In the SNSM there are a series of agents (see table 3.1) that affect, in several ways, the indigenous peoples' process of decision-making regarding development in their territories. Some of these agents' actions have been analyzed earlier. However, I want to call attention to the presence of the guerrilla and the paramilitary forces, which through violence have generated displacement and massacres of indigenous groups as well as peasants and urban peoples. There are also politicians and public figures whose perceptions of indigenous peoples generate conflicts that affect indigenous peoples' ability to govern their own lives and territories.

Guerrilla and Paramilitary

The guerrilla and paramilitaries have affected all indigenous peoples' lives in Colombia. In the middle of the 1980s the Revolutionary Armed Forces of Colombia (FARC) entered the SNSM. They settled down in certain river basis, and in the 1990s they consolidated their control of the western slope of the SNSM. Subsequently, groups of the National Liberation Army (ELN) arrived. These two armed forces have unfolded a series of actions against infrastructure and confrontations with the national military force. Further complicating the situation are the self-defense or paramilitary forces which originated from the armed marijuana growers who appeared in the 1960s. These groups have consolidated their power through the financial support of drug traffickers, cattle ranchers, banana growers and politicians. They have settled the north slope of the SNSM.

These agents have generated social, political and economic instability in the indigenous territories, which they use as temporary bases or avenues to transport illegal crops and arms, placing indigenous groups in a vulnerable position, even though they are neutral in the conflict. Likewise, they have started to pressure indigenous leaders, some of whom have been marked for death and killed due to their political activities. However, it is also important to note that there are some indigenous persons who, in a personal manner, are involved in these activities.

Politicians and Public Figures

In general, writers, politicians and intellectuals in the region have specific conceptions of the province and how indigenous peoples should be incorporated into regional programs. Most of these agents receive conceptual and academic support to promote governmental policies. Their conceptions usually conform to the general framework for managing territory and natural resources that employs national and international policies for sustainable development. However, as noted above, these regional programs and policies limit indigenous peoples' autonomy and governability.

In Colombia, each city has to construct in the next ten years a plan for sustainable development called a Territorial Ordering Plan (Plan de Ordenamiento Territorial-POT) in order to define and develop its area of influence. In the case of the SNSM, there are various plans. However, the plan that has the most influence in the indigenous peoples' territories is Santa Marta's Territorial Ordering Plan.

This POT is very interesting because it reproduces the text "*The Historical Process of Occupation of the Santa Marta Area: Elements for Its*

Rural Territorial Ordering Analysis" written by Antonio Navarro Hernández (2001). The author states that the occupation of the rural territory (SNSM) of the District of Santa Marta started in the 14th Century, and that indigenous peoples helped to destroy the environment of the SNSM due to their excessive use of slash-and-burn agriculture, territorial dispersion, the maladjustment of their social organization with their economic system, and the influence of new technologies such as metal axes.

The author also indicates that indigenous peoples have taken political actions to recover their ancestral territory, and that the state has adopted a simplistic position of expanding *resguardos* lands within the political administrative limits of the District of Santa Marta, and that it justifies this action under "the supposed millenary conservation tradition" of indigenous peoples. He argues that this approach ignores the reality existing in the *resguardos* where agricultural practices and cultural activities endanger the existence of species (endemic and others) of high biological value. He concludes that because of the state's decisions, 53% of the Santa Marta's territory does not have any governability under the District Administration, which has created an expectation among indigenous peoples that they will control the territory when they form an ETI.

The author's arguments are important because they appear in the Santa Marta's POT that addresses the issues of autonomy and governability that confront the indigenous peoples' self-determination. This is an example not only of contradiction among the state's institutions but also the role the local intellectual can play in the process of influencing the balance of power and the prioritization of issues in the relationship between regional and national policies and institutions. In addition, it clearly exemplifies a concern for biodiversity and bioprospecting derived from the discourses of environmental conservation and economic development.

On the one hand, the Ministry of Environment through its participation policy wants to establish relationships of equity and recognition of indigenous peoples' culture and rights. On the other, the policies of the Santa Marta's municipality propose to implement a plan that focuses on developing the city. In this plan, indigenous peoples are considered landowners. Santa Marta's POT suggests that indigenous peoples move from some basins with the purpose of conserving water resources. Finally, it proposes that eco-tourism and ethno-tourism activities should be developed in indigenous territories in order to conserve the cultural and environmental heritage of indigenous peoples. These proposals contradict the whole process and the policies that the CTC is trying to articulate and implement at all levels of government institutions, specifically the notion that the value of indigenous culture and practices does not depend on their usefulness to nonindigenous interests.

As we have seen, however, these contradictory visions are dispersed throughout the same complex of environmental discourses. The expectation of the officials of the city of Santa Marta is that environmentalism includes development activities such as tourism. Therefore, indigenous peoples' environmental and cultural politics become an impediment to their conception of environmentalism and indigenous peoples are designated enemies of "progress."

Governmental and nongovernmental institutions consider agreement, autonomy and participation of indigenous peoples as basic; however, it has been difficult to implement these principles in a practical way within environmental negotiations. As is mentioned in the agreements, although the government has established processes that facilitate negotiation in the SNSM, these are based on western concepts that incorporate indigenous peoples in a logic of development or preservation that effectively isolate indigenous peoples' knowledges and values from the national political processes in which they are supposed to participate.

Emergence of New Notions of Nature

The emergence of environmental awareness has generated new concerns about the relationship between humans and nonhumans as well as new concerns related to the best way to use natural resources (sustainable development). Environmental discourses employ various notions of nature. The monist notion promotes environmental discourses that seek human unity and interrelation with nature, while the dualist notion of nature as separated from society promotes notions of biodiversity and sustainable development (see chapter 4). The emergence of the idea of biodiversity has also generated the notion of genetic resources. This notion generally appears in two forms in environmental discourses: as a "natural thing" and as a "constructed thing."

These two conceptions seem to be contradictory and different, but they have similar implications for indigenous peoples. If genetic resources are "natural things" belonging to humanity, then indigenous peoples have to share these natural resources. If genetic resources are socially constructed as belonging to local people, then indigenous peoples can sell their resources as a commodity. From the western point of view, the recognition of indigenous peoples' rights is not only an alternative system for managing and developing natural resources in a more productive way, but also a system for transforming indigenous "traditions" into commodities or items of interest to social sciences. This approach to managing biodiversity, genetic resources and culture resembles the colonial processes of making inventories and collections both practically and in its underlying assumption that terra incognita and its peoples should be placed under the control of civilization for their own protection and for the benefit of the "humanity" civilization represents.

The Preservation of Nature versus Selling Nature

The protection of nature and the search for ways to use it without degrading it have become global priorities. Areas considered better conserved and with more biodiversity are thus perceived as global heritage sites. The activities of different global agents have focused on the idea of conservation as a concern for all humans. However, another fundamental interest of preservationists is the knowledge of indigenous peoples that can save humanity from global environmental destruction, as is evident to them in the fact that indigenous peoples most often live (and have lived for long periods of time) in the last areas of significant biodiversity.

However, this global need for conservation and natural balance expressed in developed industrial countries makes it evident that the destruction imposed by the markets of developed countries has surpassed all the limits of natural balance. For this reason, the idea of conserving these areas while exploiting them in a sustainable way for industrial use has become the new paradigm (and paradox) of western consumer societies. This western "solution" has also been the principal source of conflicts with local indigenous populations and the principal motive for the formation of the new social movements responsible for the creation of indigenous peoples' ecological identities and their emergence as powerful new agents within the global political context.

Some scholars (Varese 1996a, Grunberg 1995, Arvelo-Jiménez 1995, Conklin 2002) note that this coalition (indigenous/environmental movements), although strategic, has negative implications for indigenous autonomy within their territories and over their resources. The sustainable development projects that have been introduced in "third-world" countries have imposed a global management of natural resources that has not significantly considered indigenous peoples' practices and strategies. Moreover, some environmental movements of biocentric perspective promote the preservation of wilderness without considering the indigenous peoples use of their own territories at all.

In a parallel manner, the economic interest in biodiversity and the lack of regulation of the article 8j of CBD, at national and international level, is also a threat to indigenous peoples' autonomy and their territories, especially in places designated protected areas or areas of great biodiversity. Likewise, this approach introduces them into a western logic whose conceptions of individuality and the economic value of natural resources displace indigenous conceptions of the integral spiritual relationship of nature and humanity. Lorenzo Muelas (1998) has criticized these processes and highlighted the implications they have (especially those the CBD has had) on indigenous peoples' self-determination by regulating in a western way relationships that they have maintained with nature for centuries.

Separation of Indigenous Peoples' Rights and Environmental Rights

Indigenous peoples' political and environmental rights have been ratified in international and national contexts in a parallel way. In fact, international environmental and indigenous peoples' laws have become totally interrelated on paper. However, the development and the implementation of these rights have been slow and sometimes ignored. Juridical analyses show that the basic notions that sustain the rights of indigenous peoples are not applied fully because the indigenous are not recognized completely as sovereign *peoples,* especially with regard to control over their territories. Correspondingly, indigenous peoples use their own laws to restrict western efforts to control and use the environment based on western laws and modern notions of sustainable development that violate indigenous peoples' principles of the care of the environment (Flórez 2001).

In the case of the CBD, indigenous peoples are not recognized as sovereign *peoples,* neither in their rights to self-determination and autonomy over their territories and resources, nor in their collective intellectual property rights in their knowledges. The CBD recognizes the right of free and prior informed consent in relation to the states, but it is not clear if this right also belongs to indigenous peoples. The CBD has several other implicit and vaguely stated ideas regarding the "relevance" of indigenous traditional knowledges, innovations and practices pertinent to conservation and sustainable use of biodiversity. Therefore, it is not clear what the CBD means by "relevance" or what is considered relevant in "traditional" styles of life. Finally, the CBD does not recognize indigenous peoples' juridical systems.

The UN Permanent Forum on Indigenous Issues does not even include in its name recognition of the indigenous as sovereign *peoples*. Likewise, it does not have the power to generate recommendations to members of UN. The indigenous peoples' members of the forum participate as individual management experts rather than political representatives of indigenous peoples (COICA 2001).

However, it is important to note that in the Political Declaration of Johannesburg (2002), in their paragraph 22 (bis), and thanks to the pressure of indigenous peoples, a text was included that says: "We reaffirm the vital role of indigenous peoples in sustainable development." This was the first time that the term, *Indigenous Peoples,* had been used in this context.

In the Colombian context, indigenous peoples have rights over their territories and resources. In the same way, the ILO-169, or Law 21, of 1991, propose previous consultation as a fundamental right of indigenous peoples in their efforts to determine their own priorities and as an instrument to defend their cultural integrity, political participation and autonomy. However, the decisions that indigenous peoples make through previous consultation cannot determine any environmental authority's final decision. In the last declaration of ONIC, indigenous peoples stated that development projects are forced onto their territories, and consequently they are losing their rights of political autonomy and participation as well as control (property rights over resources) in their territories.

THE PURSUIT OF ECO AND ETHNO-ACTORS

> The natural ingredients supplier Unigen Pharmaceuticals (UPI) said this Week it is to collaborate with consultants Botanical Liaisons to discover new plants for use in medicines. The companies will look for new products in countries including Belize, Cameroon, Peru, Tanzania and the United States. Botanical Liaisons will identify, collect and supply medicinal plants with voucher specimens and traditional information to UPI for research and product development. UPI will then screen the plants for biological activities, isolate and purify the active agents to make them homogeneous and work on developing drug leads and commercial products. [. . .] Botanical Liaisons adapts traditional intellectual property into social sound projects, which it claims are mutually beneficial to native communities and entrepreneurs in the natural products marketplace. The company will also benefit from UPI's research program, which it is hoped will lead to economic and environmental development of successful botanicals. UPI and Botanical Liaisons state that they comply with the terms of the 'Convention on Biological Diversity' that allows sharing and recognition of traditional knowledge. Both companies are also committed to conservation and sustainable use of medicinal plants indigenous to the source countries. (FoodPro-

> ducttionDaily.com. Breaking News on food Processing and Plants. Search is on for new medicinal plants 29/1/23. http://www.foodproductiondaily.com/news/news.asp?id=2126)

The global processes of environmental protection and economic development that focus on biodiversity and genetic resources require that the relationship between indigenous peoples and environmentalism be analyzed according to new relationships of power/knowledge. These conservationist ideas and economic processes have generated a new conceptualization of nature as a human heritage and an eco-commodity.

Following Gupta (1998), I argue that the West's interest in indigenous peoples is ambivalent. Such interest is deeply connected to the neocolonial and nationalist imaginings of indigenous peoples' traditions that pervade development projects. Gupta (1998) also notes how indigenous knowledges may also be affected (usually negatively) by capitalist processes that introduce them into circuits of capitalist production and consumption through tourism (eco and ethnic), the search for raw materials for industry and medicine (timber, plants, petroleum and minerals), the quest for new genetic resources, and actual processes of colonization on "low-density frontiers" in the nations of the so-called "third world." He explains that the increasing attention paid to indigenous peoples' knowledges among nonindigenous people is based on the recognition of their economic value to the nonindigenous. Moreover, he shows how the inclusion of indigenous peoples' knowledges has become an important legitimizing strategy of transnational ecological, agronomic and development discourses, especially when indigenous peoples serve as informants to these projects. In addition, the inclusion of indigenous knowledges and participants in bioprospecting processes has reduced the costs and increased the benefits for transnational pharmaceutical corporations.

We can also see examples of such situations in the case of indigenous peoples of the SNSM and their political strategies. Although they have gained political power within the national context, and the CTC in particular has gained power to negotiate with the World Bank and the GEF project, these negotiations continue to be framed in western economic terms rather than indigenous cultural terms. GEF has therefore encountered indigenous opposition to its economic development processes.

> Almost inevitably therefore, many GEF project-affected peoples call for its funding to be halted, reconsidered, redirected, while others speak of an arrogant, arbitrary and slow-moving bureaucracy embodying (behind the scenes) 'an enormous con'

> promising vastly more than it can deliver, 'greenwash' for the World Bank's environmentally destructive developments in forestry, dams, energy, transport, etc. and financial 'sweetener' for international lending and thus, third world debt. In this context, the politics of influence in and around the GEF can shine a light on conservation and colonialism, capitalism and complexity, compromise, co-option and commodification in a rapidly transforming world (Young 2002:1).

In the SNSM, indigenous peoples' dynamics and interrelations have shown us how the ideas of collective identity, territory and sovereignty have been reinforced in terms of a nation-state conception that fosters a new kind of 'micro' nation-state able to negotiate with transnational corporations and influence transnational policies, and indigenous peoples are indeed negotiating their participation in the GEF project through a direct relationship with the World Bank and with an environmental NGO (FPSNSM) that is implementing national and global environmental policies instead of the state. National institutions such as the Department of Planning and the National Natural Parks' Unit are also enabling such negotiations. Yet, as suggested above, these apparent instances of indigenous empowerment come at a price.

The global environmental policies that began to appear at the beginning of the 1980s were the results of increased environmental awareness, the environmental crisis and the search for new economic alternatives to the modern or progressive western model of development. Most of these policies addressed the environment in terms of devising "sustainable" models of economic development of natural resources to resolve long-term social and economic problems of the inhabitants of the designated regions (see chapter 4).

In this articulation of the environment and development new scenarios appeared that redefined natural resources within a conservation logic, as in the case of the World Conservation Strategy (UICN 1980), and that promoted the view that protected areas could contribute to the economic development of local inhabitants. Furthermore, this view generated more proposals among institutions and programs such as the Environmental Program of the United Nations (UNEP) and the World Park Congress (1982) that in a more explicit manner linked conservation, people and development.

The linkage of environmental conservation, indigenous peoples and development generated several outcomes, among the most important in the present context, being eco-tourism programs. These programs operate under the premise that poverty can be diminished through an economic exchange between "developed-country" tourists and "nondeveloped" country inhabitants who have environmental resources of ecologically symbolic interest to such tourists. The result is the introduction of new territories with biological diversity to global processes of control under the logic of "global heritage" and the economic valuation of biodiversity as a new global commodity: eco-products in a green market.

In sum, the opening of new territories with biodiversity under this mode of the sustainable-development rubric introduced not only territories but peoples as consumer items in these green markets, generating new products and services under the eco-product criteria and developing entire new industries such as eco-tourism.

Eco-and Ethno-Tourism

The emergence of eco-tourism in the 1980s occurred for various reasons: the romantic search for natural areas without human intervention; the enjoyment of nature under a "nonconsuming" premise as an affirmation of environmental awareness; and the contribution to the site's conservation through the profits from tourism realized by local inhabitants and the consequent reinforcement of their interest in its conservation. Local people are thus encouraged to develop eco-tourism through environment and "cultural attractions" with the expectation that such a business can provide a solution to their social and economic problems (Campos 2002).

The western desire for natural or pristine sites has generated eco-tourism in national protected areas insofar as they are perceived to be wild and untouched spaces. However, in most countries, these areas belong to ethnic groups and indigenous peoples involved in two parallel processes: a struggle for the recognition of indigenous peoples' and local communities' territorial rights, and new globalization processes promoting the implementation of sustainable development projects that make those territories commodities in the national and global eco-markets. The meeting of these processes has caused many conflicts and contradictions.

Eco-tourism has had an unusual development and an important role within the agendas of national and international development institutions as an economic strategy to promote a national and foreign tourist industry. Within eco-tourism, and based on the fact that these natural places belong to "real people," the *ecological native* began to assume an important role in the marketing of special destinations for "ethno-tourism." These ethnic and eco-touristic images are produced and disseminated through the mass media as part of national development policies (see chapter 6).

For example, in Colombia, the Ministry of Development established how state policies of eco-tourism should be implemented. In 1994, the Ministry of the Environment along with the Ministry of Development began preparation of national eco-tourism policies to conserve and develop zones that had any potential attraction and to increase the knowledge and awareness in local, regional and national settings of their potential for touristic development. In 1999, the Ministry of development published the "Legal System of Tourism in Colombia" (Régimen Jurídico del Turismo en Colombia-RJTC) and the "Competitive Study of the Tourism Sector" (Estudio de Competitividad del Sector Turistico-ECST). This legislation established the state's position on environmental policy in relation to these new eco-industries.

First, it indicated that eco-tourism is "that specialized and directed form of tourism developed in areas with a special attraction and framed within a human sustainable development parameter. Ecotourism is a controlled and directed activity that produces a minimum impact on natural ecosystems, respects cultural heritage, and educates and sensitizes the involved agents about the significance of conserving nature" (RJTC 1999).

It also emphasized that ethno-tourism is "a specialized and directed tourism that is carried out in territories of ethnic groups with cultural, educational and recreational purposes that allows learning about cultural values, forms of life, environmental management, ethnic groups' customs, as well as learning from aspects of their history." This legislation establishes that the Ministry of Environment, in coordination with the Ministry of Economic Development, should administer, protect, preserve and regulate the use of National Parks within the framework of the sustainable development planning. As Law 99 dated December 22, 1993 states, the Ministry of Environment was created "as an organization ruling environmental management and renewable natural resources, in charge of promoting a respectful and Harmonious relationship between man and nature and defining policies and regulations to which the recovery, conservation, protection, management, use and exploitation of the natural resources and of the national environment would be subject to, in order to assure sustainable development." Likewise, it indicated that the state should promote eco-tourism and ethno-tourism and that these should be included within tourist development plans prepared by the Ministry of Development. Moreover, regional projects should be coordinated by autonomous corporations or national sustainable development programs to favor nature's protection and conservation programs.

The implicit argument supported by the law is that nature's preservation should be under the parameters of sustainable development and under the coordination of the Ministry of Environment and the Ministry of Economic Development. Nature under the state view is an economic priority and a national heritage that should be exploited and maintained by guaranteeing that resources can be renewed and the development programs can be carried out within protected zones, including indigenous peoples' territories.

The "Competitive Study of the Tourism Sector" (Estudio de Competitividad del Sector Turistico-ECST) proposes that in the long term, the national tourism strategy should be focused on eco-tourism as a promising economic prospect for the country, especially since Colombia has natural resources with great value in the world market and eco-tourism offers work to local people (such strategy also serves a national, social, and political goal of maintaining the presence of the state in marginal zones). According to the study, the eco-tourist wants a less-developed and better-conserved nature as close to a pristine state as possible. The nation-state thus has an interest in guaranteeing this "natural" state.

According to these documents, natural parks are also important from a touristic perspective because many exotic indigenous peoples' communities exist in them (in Colombia there are 18 natural parks that are superimposed on indigenous peoples' territories), which provide the tourists with the chance to learn in a "live manner about indigenous peoples' ancestral culture." For this reason eco-tourism and ethno-tourism have developed jointly. Eco-tourism is presented as one of the few possibilities to generate income in "marginal" and "poor" communities. The studies go on to suggest the creation of eco-shops and eco-shows to employ local people and highlight their ethnic characteristics.

From the state's perspective, the environment is an economic development factor and tourism a potential aid in preserving natural resources. Eco-tourism is consistent with the government's policy to discourage activities that are detrimental to the country's progress, such as the inadequate exploitation of natural resources, which in this case include the natural environment and its "exotic" peoples. Consequently, it promotes the Strategic Eco-system Protection Program (Programa de Protección de Ecosistemas Estratégicos) for those eco-systems that "offer the necessary support for the country's economic development, through environmental assets and services and constitute an essential instrument for the sustainable development that assures productive processes' continuity, environmental equilibrium, risk prevention and biodiversity's conservation" (ECST).

In all these programs what is evident is an economic vision that converts nature and indigenous peoples into objects for consumption through state planning. The idea that supports the programs is one of a pristine nature that should be conserved as such and used as a generator of jobs and profits. Indigenous peoples, because they live in those pristine sites, are part of them, and at the same time they are making the products that have to be "sold" to tourists whose consumption (and consequent redistribution of money) can "improve" indigenous peoples' conditions. To promote this relationship, the government has initiated a strategic manipulation of views of nature and indigenous peoples through the mass media (see chapter 6) that provide the eco-market with idealized images to encourage visitors to explore these places. For example, one document suggests the necessity of building in the SNSM a luxurious eco-hotel in the Tayrona national park and an eco-hotel for the Lost City despite the fact that such developments would destroy the environment and culture that provides the rationale for the developments. Indigenous peoples and ecosystems thus constitute an economic factor in the government's national tourism programs rather than a truly cultural or ecological one.

Likewise in other regions of the country, indigenous peoples' communities are promoted as eco-tourism destinations (especially in the Amazonas, Chocó and Guajira regions), because these are the regions thought to preserve best the pristine nature that attracts eco-tourists. In addition, they offer the exotic charge gotten from having the chance to learn about and share the indigenous peoples' lifestyle.

Evidently, biological diversity and indigenous peoples have become marketing strategies within an economic discourse. Tour packages now offer visits to indigenous peoples' communities that encourage indigenous peoples to portray themselves as exotic objects and to "exhibit" themselves as natural components within their exotic landscapes. This interaction of indigenous peoples and nature also appeals to westerners' desire for a spiritual connection to nature that can be had (bought) through contact with the "wise" *ecological natives* in their pristine natural setting.

The opening of the SNSM to ecological and economic processes like eco- and ethno-tourism is relatively new. Nonetheless, it may also be seen as yet another economic process and wave of colonization among the many that have introduced the SNSM to national and global markets over several centuries. What is truly new in this latest effort to develop the SNSM is the nature of the market and consumer desires that motivate it: green markets that produce organic/eco-products and ethnic- and eco-tourism (development) for people who want to escape the effects of development in their own countries. In the case of the SNSM, the government's efforts to attract eco-tourists center on the National Natural Parks Tayrona and Sierra Nevada de Santa Marta. The national policy related to eco-tourism there states:

> The true potential of the region is that it may get to displace other tourist sites on the coast. According to tourism companies, there you can find the best natural landscapes and the most important ethnographic and archeological redoubts of the country. In this manner the strategy shall focus on improving hotel capacity in these sites, preventing deforestation, arrival of new settlers, and accumulation of nondegradable wastes, also offering eco-tourism packages that include beaches, going up the Sierra and the visit to the Lost City, which shall be presented as an attraction in Santa Marta: remote, mysterious, and with a great mythic load.

In the city of Santa Marta, major local policies conform to recommendations of national and international eco-tourism policies, and newspapers dedicate articles to promoting the SNSM as an ecological paradise and cultural fantasy: a perfect advertising strategy to attract tourists seeking to escape the world of profits, progress and development in which they live.

Santa Marta's eco-tourism sites are centered in the Tayrona Park and in the Lost City. These sites are described as having traces of the wisdom of indigenous cultures, although which cultures remains vague since several descriptions make no distinction between the Kogui people and the Tayrona people (a possibly distinct pre-Hispanic indigenous culture). The architectural legacy of the latter is highlighted, and the Kogui are presented as the living heirs of the most authentic indigenous civilization that still survives in the Americas. Emphasis is also given to the Kogui's harmony with nature, and their territory is described as a magic environment, with exuberant vegetation and sacred sites from a one-thousand-year-old mysterious culture. Consequently, a trip to the Lost City or Pueblito or any other architectural remains of the Tayrona's "authentic" origin is presented as a major ethnic and ecological adventure (*El Tiempo*, May 4, 1992) to a "remote and almost inaccessible place where you have the privilege to penetrate into the fantasy of a one-thousand-year-old culture and an unbounded nature" (tourism booklet of Irotama Resort-Golf-Marina).

The nonindigenous perception of the indigenous peoples of the SNSM and their environment is largely based on a sacred character and ancestral history invented and maintained by the imaginary of eco-tourism. The imaginary (if not simply fantastic) quality of this experience is evident in the promise that the tourist will have an authentic encounter with a natural, cultural and spiritual otherness while still having time to golf in the afternoon. In another instance, tourists are told that a reconstruction of an indigenous site in Parque Masinga is an "important archeological reserve of Santa

Marta with a thick vegetation made by gallery woods where one can feel the energy and force of nature and with a pair of circular terraces rebuilt with two huts as replicas used in the past by *Mamas* and governing cabildos." In these places tourists are advised that they are entering into a sacred site that is worthy of all their respect, especially their silence in the presence nature's fragility (*El Tiempo*, March 28, 1996). Such a "virgin" nature is in itself presented as a sacred remnant of Eden, much like the Amazonian region (Slater 2002).

Eco-products

Eco and ethno-tourism areas also support eco-markets for tourists. In the case of SNSM, organic products associated with the Kogui's spiritual world have flooded local and foreign markets. Among these products organic coffee production is important, particularly the coffee processed and traded in England called "Sanctuary Organic™", a gourmet brand promoted with a picture of a Kogui man and with a text that reads "The legendary coffee of the Kogui indigenous peoples." The label discusses the SNSM and how the coffee gets its name due to the unique conditions in which it grows: in the midst of the tropical rainforest in a natural conservation region where the Kogui use centuries-old agricultural processes. Finally, it mentions the Kogui's ancestral history and their respect for the environment, and it also explains that a portion of the profit will go to the producer (10%) and an extra percentage (5%) to local communities to help them re-purchase their sacred territories.

Ethno-tourism sites also frequently offer eco-products or souvenirs for the consumption of tourists. These products have to be made from local eco-safe materials (which is a contradiction insofar as the environment is subject to increased extraction activities to produce them for tourists), and hand made, especially by women or elders, which signifies spirituality and a special contact with nature. These eco-products usually are representations of animals and spirits of nature represented on earrings or necklaces made in wood or other natural materials. These products eventually become industrial productions: the quantity and quality that tourists demand eventually detaches the object from its cultural context and converts it into a pure commodity.

MARKETING KNOWLEDGES

> The discourse about indigenous ecological wisdom has shifted away from claims about the superiority of specific indigenous

> resources management practices and toward claims about the value of indigenous knowledge that transcends the limitations of Western scientific knowledge. Instead of focusing on what Native communities actually *do,* this discourse highlights the potential value of what indigenous individuals may *know.* Whereas Natives peoples formerly were positioned as guardians of the forest itself, now they are positioned as guardians of *knowledge* of the forest. (Conklin 2002:1056)

Indigenous knowledges have been named in different ways: native, aboriginal, folk, traditional, local, peoples' science, and rural peoples' knowledge, among others (Ellen, Parkes and Bicker 2000, Long Martello 2001, Brush and Stabinsky 1996). Since colonial times, indigenous knowledges have played an important role in the development of medicines, conservation ideas and western science (Grove 1995, Nieto 2000, Ellen, Parkes and Bicker 2000). However, with the expansion of western science and the increasing intellectual and political authority of scientific experts, knowledges that did not conform to the nature/culture dualisms at the basis of western science were ignored, marginalized, undervalued and persecuted (Ellen, Parkes and Bicker 2000). More recently, however, indigenous peoples' knowledges have acquired new, positive connotations in light of new processes of valuation of biodiversity that locate them in national and transnational contexts.

These previously-ignored knowledges now enter into the discourse of biodiversity for different reasons: the emergence of environmental awareness; the cultural politics of indigenous movements; the repositioning of indigenous representations; the recognition of indigenous practices within environmental and developmental discourses; the change in conceptualizations of nature; the necessity of constructing a new society due to the crisis of economic development; the epistemological shift in natural and social sciences concerning notions of nature; and the introduction of the territories and resources of indigenous peoples into commercial circuits, among others.

However, I want to focus here on the economic reasons for the recognition of these knowledges. The relationship between the new economic valuation of knowledges and indigenous peoples is enunciated in different global policies such as the CBD. In Colombia, environmental policies, such as Law 99, establish the importance of indigenous peoples' knowledge in the following manner: "the Ministry of Environment and national scientific institutes will promote the development and dissemination of knowledges, values and technologies related to the management of the environments of indigenous peoples and ethnic groups." And the reason for this interest is that indigenous knowledge has economic value insofar as it reduces costs and legitimizes and leads to the discovery of new products that can be patented.

Reduction of Costs

The inclusion of indigenous peoples' knowledges in bioprospecting processes reduces costs and increases the economic benefits of transnational pharmaceutical corporations. For example, researchers value indigenous peoples detailed knowledge of environments that contain scarce "global commodities" that face environmental destruction, and because indigenous peoples' knowledge of their territories facilitates the bioprospecting acitivities, and in this manner indigenous peoples' knowledges are now themselves purchased as products or services in this new ecological free market (McAfee 1999, Gupta 1998, Escobar 1999, Sachs 1999). In other cases, local people (indigenous, peasants, etc.) are hired as biologists or researchers who administer western science to collect data and systematize it. It is important to ask the implications of such development strategies, and how institutions and NGOs can recognize local participants' intellectual property rights and share the benefits of them. Under current circumstances, their knowledges achieve most of their monetary value after they have been transferred to NGOs and research institutions and transformed into their proprietary knowledge.

Indigenous peoples' knowledges and their importance to ecological conservation have been recognized by the CBD, however the economic value of these knowledges has transformed their cultural and ecological significance to economic terms. These conceptual changes can be seen in how indigenous knowledges' cultural value and environmental significance have been transformed in discussions of the WIPO and Word Trade Organization-WTO that address indigenous issues in terms of individual property rights and their place in global development and trade. The following document represents the kind of indigenous participation that most readily receives recognition and support from western interests (one that conforms to the commercial logic of the West).

> Vancouver/Kenora, January 30th, 2003: The Indigenous Network on Economies and Trade (INET) today released its official amicus curiae submission to the World Trade Organization (WTO) regarding the final determination in the Softwood lumber dispute. In an unprecedented move, the WTO accepted the first ever substantive Aboriginal submission filed with the WTO

> by member nations of INET, who are confident that this second more comprehensive brief will also be considered in the WTO deliberations. (UBCIC's Protecting Knowledge Conference site: http://www.ubcic.bc.ca/protect.htm List information: http://groups.yahoo.com/group/protecting_knowledge)

Legitimization of Processes

The inclusion of indigenous peoples' knowledges has been important for transnational development due to its practical benefits and due to the fact that indigenous informants legitimize these programs by making them appear mutual rather than unilateral (Gupta 1998). Also, through this strategy, every day more indigenous peoples are "qualified" to speak the language of development. In fact, the World Bank pays for the training of indigenous peoples in strategies of development.

In most of the cases the supposed dialogue between different knowledges in fact introduces indigenous peoples to the logic of environmental programs and development in a manner that ignores their knowledges and encourages local support for the new western logic. Most international economic donations for environmental programs and development have as a basic requirement "local participation," and the result is most often the indoctrination of indigenous peoples in the discourses, values and goals of development (or the "civilization" of indigenous peoples).

Building Commodities

> The World Intellectual Property Organization (WIPO), the Geneva-based body that promotes intellectual property rights, is keen to establish databases in which indigenous groups would record their cultural knowledge. Patent examiners would be encouraged to check a country's databases to see if a new idea, such as a plant-based medicine, is part of the traditional knowledge of that nation. But delegates at the World's Indigenous Peoples conference, held on 16–19 October in Kelowna, British Columbia, described how some groups are concerned that the databases could be used to exploit their cultural heritage (Dalton 2002:866).

Indigenous knowledges are valued because of their integral relation to nature, yet as these knowledges become a product in themselves, a conceptual change occurs that separates them from their integral relationship with territory. Knowledge becomes a product independent of its culture and environment of origin; that is to say it becomes a commodity that can be patented and marketed independently from the sociocultural and ecological contexts of its production. The displacement of politics for the creation of a standard way to manage and control these knowledges through databases and formats decontextualizes them and relegates cultural particularities and differences to a secondary plane, thus making them unavailable to the discussion of the power relationships or politics involved in their use.

Indigenous knowledges first appeared as a topic of in the discussions of the CBD regarding conservation strategies, but now they have become a topic in the discussions of intellectual property rights in the WIPO. Indigenous peoples' knowledges have acquired a dual and contradictory status. One is based on their relationship to the conservation of biodiversity as a component in the struggle to strengthen indigenous cultural identity and property rights. The other is based on western economic interests that require the expropriation and transformation of indigenous knowledge of biodiversity as a component of development. Nevertheless, western knowledge is the one that decides the processes and goals indigenous knowledge will serve. Therefore, it decides how and why to conserve territories with biodiversity, thus following a well-known course toward the domination and negation of the cultural differences that made indigenous knowledges valuable in the first place. As Escobar (1999) argues, sustainable development promotes "the belief that it should be—once again—the benevolent hand of the Westerner that saves the earth [. . .] westerners are those who must reconcile humanity with nature. Only in a second instance are other communities invited to share their 'traditional knowledge'" (84).

For example, the West takes an instrumental and self-interested view of indigenous peoples' knowledges of flora, as in the case of the Barasana people who can identify all species of trees in their territories without referring to flowers and fruits, or the Inga people of the Putumayo with their knowledges of medicinal plants as alternatives for treating terminal diseases. In this sense, knowledges that indigenous peoples possess are taken into account for their economic value or utility in medicine or scientific research, a vision that implies an understanding of indigenous peoples' knowledges in terms of property, land use and the exploitation of nature. The monetary rewards of genetic patents, then, raise a whole set of new conflicts.

The problem is apparent in who receives the health and financial benefits of indigenous knowledges. For example, the world market in medicinal plants and their derived products is estimated at US$20 trillion, and the herbal-medicine market grows ten percent annually in Europe and North

America. Although Colombia has a great potential in genetic resources that can be exploited industrially to generate sustainable economic development in indigenous communities, the global tendency has been toward patenting products without considering indigenous peoples' collective property rights (biopiracy). The interests of indigenous peoples and Colombians in general are secondary to the interests of western peoples ("investors") in their own health and wealth. As Escobar (1999) states, "The protection of intellectual property of living matter is being promoted by international entities not as a form of protecting communities of the Third World, but to assure its privatization and exploitation for capital" (95).

Despite their environmental good intentions, western peoples urgent desire to preserve, conserve and protect nature ultimately finds expression in a capitalist conception that sees in nature and its associated knowledges an economic potential: a commodity that can offer a variety of products and services for profit. In this context, the protection of the environment depends on its value in the context of development or sustainable development. Eco-products then appear under this eco-capitalist rubric with the willing or unwilling cooperation of indigenous communities.

Another example is the case of an indigenous peoples' community in the Amazon that makes ecological paper through manual processes employing organic waste and that also makes elaborate fabrics from *moriche* palm. The production of these new eco-products appear to offer indigenous peoples alternative forms of development designed for their benefit, but the reality is that their lives and cultural traditions are being assimilated and changed by the values of an international economic system and its eco-consumers demands for genuine "eco-products." Environmental services are also promoted in this manner. The sale of the "clean air" in Colombia's indigenous territories with great forest biodiversity is being proposed to foreign investors as a means to diminish the contamination of the planet by creating carbon dioxide sinks to clean the pollution generated by developed countries (*El Espectator*, May 24, 2000).

Colombia's national environmental policies are greatly influenced by the demands of global environmental policies that emphasize sustainable development and the profits that can be generated from the Colombian environment and its high biodiversity. From this point of view, nature is a resource that should be exploited, and this necessitates implementing a national policy related to genetic resources, such as that of the Von Humboldt Institute. That policy proposes to link the economic development of genetic resources to indigenous knowledges in ways that will "protect traditional knowledge" through indigenous empowerment in the market place.

Again, the assumption is that indigenous knowledges and cultures have little or no value or power until they have been integrated into the discourse and practices of profit-driven development programs.

Bringing Traditions to Modernity

The consumption of indigenous peoples' environmental knowledges demands that indigenous peoples maintain their "indigenouness" and "difference" while at the same time making them available as commodities. On the one hand, this gives indigenous peoples motivation and support for protecting or recovering their traditions. On the other, these traditions have to be maintained according to certain western conceptions of *ecological native* life and tradition, but in such a way that they are available to non-traditional western values and interests. As a result, the maintenance or reinvention of these traditions constitutes a response to western problems rather than those of indigenous peoples, and so western interest in these traditions is selective. For example, indigenous knowledges related to medicinal plants are valued because they fit into bioprospecting processes that serve western health priorities, and this sort of interest, rather than an inherent respect for indigenous cultures, peoples or their territories, motivates most western efforts to promote indigenous cultural identities and territorial rights.

FINAL REMARKS: FACING ECO-GOVERNMENTALITY

To the extent that debate and conflict continues regarding the meaning and use of the figure of the *ecological native* in local, national and transnational discourses of all sorts, there is reason to believe that indigenous cultures and their peoples can modify and even change the underlying values and desires that drive western (or westernized) cultures and their peoples to lead the world toward a future of ever-greater environmental danger and difficulty. We will all be losers if we allow current western values and practices to achieve a truly hegemonic status by refusing to consider and act upon alternatives to them. The *ecological native* is a figure of a troubling past, an uncertain present and a potentially better future. If those peoples represented by the *ecological native* win our attention and respect, we all stand a chance of becoming winners.

The Best Last Hope

The interrelation of indigenous peoples and environmentalism has come with contradiction and paradox. Nevertheless, global eco-governmentality (with its practices, discourses and representations) has enabled indigenous peoples' movements to create arenas for cultural differences and rights within national and transnational political arenas. Indigenous peoples have won a political position from which they can demand and, sometimes, realize political, civil, social and cultural rights. These successes can be seen in national and transnational policies and laws that do recognize the value of indigenous peoples' territories, ecological practices, and knowledges. Moreover, the terms of this recognition offer indigenous peoples options. They now have sufficient political, economic and symbolic capital to begin the 21st century as politicians, capitalists or entrepreneurs within a new free eco-market defined in western terms. Or in other cases they have a sufficient quantity of such capital to effectively disengage from western values and practices and thus remain relatively disengaged from western concerns and desires.

And what is the problem with those options? The problem is that "we"—the modern, capitalist West—*need ecological natives.* We need them to be uncontaminated by capitalist markets. We need them to be engaged with us as our subjects, subalterns, opponents, or as our colonized, underdeveloped, or oppressed others. They must not become us—self-interested strategists, rational investors, rights-bearing western individuals, and modern disciplined bodies—and they cannot disengage from our problems and leave us to our own devices. In some form, they must be "our" others.

They have to be "our" utopian reality. They have to be the warriors of "our" inner conflicts. They have to allow us to think, believe and dream about rights, alternative relations with nature, alternative developments, political changes, new societies, peaceful worlds, social justice, equal power relationships, counter-govermentalities, counter-globalizations, alternative modernities and perfect democracies. They have to help us criticize the society that we have made, to construct sophisticated intellectual apparatuses to reflect on ourselves and to deconstruct our history and create alternative imagined communities (truly civil societies, social movements, transnational communities, social forums, third-world networks or virtual ecological communities). They have to help us be postmoderns, postcolonials, deconstructionists, poststructuralists, transcendental historicists, skeptics about scientific knowledge, critical theorists, environmentalists, activists, organic intellectuals and dissident intellectuals. They have to transform our environmental disciplinary mechanisms into tactics of resistance. They have to become our icons of political freedom. They have to do all of this for us, while we are comfortable, secure, well-paid and safe in our houses, offices, universities, libraries, meetings, seminars, classes or resorts talking, thinking and picturing them. And we often fail to see in their imperiled differences, the differences we obscure with images of our desires, our own responsibility to change ourselves and seek solutions in our own lives for the problems we have created and that we want them to solve. They are our best last hope.

Notes

NOTES TO CHAPTER ONE

1. The third wave of democratization or the process of promoting democracies in the world since 1974 has promoted in this region the increasing politicization of indigenous peoples' movements.
2. Foucault's concept of governmentality has been used to analyze and critique environmental discourses: ecological governmentality (Rutherford 1999), environmental governmentality (Denier 1999), and environmentality (Luke 1999a). I am using the concept of eco-governmentality in a similar way; however, I address the specific historical interrelation of indigenous peoples and eco-governmentality.
3. The OGT is an indigenous organization that emerged in Colombia in 1987. Kogui, Wiwa and Arhuaco peoples who live on the northern slope of the SNSM form it.
4. The CTC is an indigenous organization formed by the four indigenous organizations of the Sierra Nevada de Santa Marta for the coordination of indigenous peoples with the state and the national society in Colombia.
5. A similar methodology has been popular in Latin America since the 1970s (Fals Borda and Rodriguez 1987, and Vasco 2002, among others), however this specific approach has been developed in relation to environmental issues. In the North American anthropological tradition, this methodological perspective has been considered as part of the postmodern approach since the 1980s (Marcus and Fischer 1986, Clifford and Marcus 1986, Behar and Gordon 1995, Gupta and Ferguson 1997).
6. This approach has been basic for political ecology since the 1980s. Specifically, it makes reference to the political ecology of actor-centered approaches. It also has relations with so-called network analysis. In the North American anthropological tradition, this methodological approach has been considered part of the postmodern perspective. Marcus (1995) calls this approach "multi-sited ethnography," and Merry (2000) names it "deterritorialized ethnography." In this text, I use the term multi-sited ethnography to refer to multiple actors and sites within the analysis of environmental concerns and indigenous peoples' relationships.

NOTES TO CHAPTER TWO

1. See Urban Greg and Joel Sherzer 1991, Findji 1992, Van Cott 1994, 2000, Avirama and Márquez 1994, Brysk 1994, 1996, 2000, Varese 1995, 1996, 1996a, 1996b, Yashar 1996, 1998, 1998a, 1999, Berraondo 1999, Nash

1997, Assies et al. 1998, König et al. 1998, Warren 1998, 1998a, Ramos 1998, Gros 1998, 2000, Nelson 1999, Bengoa 2000, Laurent 2001, Ulloa 2001, Chaves 2001, Zambrano 2001, Archila and Pardo 2001, Tilley 2002, Hale 2002, Hodgson 2002, Conklin 2002, Vasco 2002, Warren and Jackson 2002, among others.

2. In Latin America, such movements also appeared, however they were not addressed within the literature of that time.
3. See in general for Latin America: Escobar and Alvarez (eds.), 1992, Alvarez, Dagnino and Escobar (eds.), (1998), Archila and Pardo (eds.), (2001). Indigenous peoples' movements are described in the next section.
4. See Ramírez (2001) for information about peasants' movements in Colombia.
5. I want to use the meaning of this word in the sense of Warren and Jackson (2002 a) who said "Note that 'indigenist' has two well-established, potentially confusing meanings. The first, more often used in reference to the policies of a state toward its indigenous peoples, indicates an integrationist position; this is particularly the case in the Spanish and Portuguese cognates. The second meaning refers to any individuals (Indian or non-Indian) or institutions in favor of indigenous rights" (36).
6. To this Pact belong the Japanese Ainu populations and indigenous peoples of Taiwan, India, Thailand, Malaysia, Indonesia, Bangladesh and Nepal.
7. ILO-Convention No.169 (1989) has been ratified by 17 countries: Argentina 03:07:2000, Bolivia 11:12:1991, Brazil 25:07:2002, Colombia 07:08:1991, Costa Rica 02:04:1993, Denmark 22:02:1996, Dominica 25:06:2002, Ecuador 15:05:1998, Fiji 03:03:1998, Guatemala 05:06:1996, Honduras 28:03:1995, Mexico 05:09:1990, Netherlands 02:02:1998, Norway 19:06:1990, Paraguay 10:08:1993, Peru 02:02:1994, and Venezuela 22:05:200.
8. Countries and dates of constitutional changes in relation to indigenous peoples' rights: Argentina (1994), Bolivia (1967, 1995), Brazil (1988, 2002), Colombia (1991), Costa Rica (1991), Ecuador (1998), El Salvador (1983, 2000), Guatemala (1985, 1993), Guyana (1970, 1996), Honduras (1982, 1999), Mexico (1992, 2001), Nicaragua (1987, 1995), Panama (1972, 1994), Paraguay (1992), Peru (1993), and Venezuela (1999) (Clavero 2003).
9. In Colombia the indigenous peoples' territories are called *resguardos* and are legal collective territories.
10. *Terrazgueros* were indigenous peoples who worked for a "new owner" of the land and received part of the harvest as a payment.
11. Some of the main local and regional indigenous organizations are: La Unión de Indígenas Jivi Sikuani del Meta y Vichada-UNUMA (1972), El Consejo Regional Indígena del Vaupés-CRIVA (1973), El Consejo Regional Indígena de Risaralda (1978), La Organización Regional Embera Wounan-OREWA (1980), El Consejo Regional Indígena del Occidente de Cladas-CRIDOC (1981), La Organización Regional Indígena del Casanare-ORIC (1983), La Organización Ingano del Sur Colombiano-ORINSUC (1985), El Consejo Regional Indígena del Medio Amazonas-CRIMA (1985), La Organización Indígena del Putumayo-OZIP (1987), La Organización Indígena Uwa del Oriente Colombiano-OIUOC (1987), El Consejo Regional Indígena del

Guaviare-CRIGUA (1989), El Cabildo Indígena del Trapecio Amazónico-CIMTRA (1989), La Unidad Indígena del Proyecto Awa-UNIPA (1991), among others.

12. The Uwa people are making a national and international protest because they want to protect their Mother Earth from petroleum exploration by the Occidental Petroleum Corporation (OXY) that is the largest U.S-based multinational company in Colombia with worldwide interests in oil exploration, among other industries. In this case, international support is expressed through a national and transnational network of environmental and human rights NGOs that has been organizing protests around the world. The Uwa people want more than to confront a politician in the U.S, they want international support from environmental and human rights organizations to denounce the Colombian government in international courts because of its violation of indigenous peoples' fundamental rights. They reject petroleum exploration near their territories. For them, it means, "to take of the blood of the Mother Earth that will produce water scarcity, the animals' extinction and the extinction of indigenous culture by the white men" (Uwa's declaration 2000).
13. Since 1993, the Embera-katio people have led a national and a transnational campaign to denounce the environmental and cultural consequences of the construction of the hydroelectric dam "Urra" on the Sinu River. At this moment they are facing violent opposition from paramilitary groups. International environmental networks, such as Global Response, Amnesty International and Survival International are denouncing this process. Their campaign has been more successful in international arenas than national arenas. The Embera-katio cause has generated a great awareness in the general public of foreign countries because political leaders have been killed and Kimy Pernia, who is one of the most important Embera-katio leaders, disappeared.
14. Stavenhagen (1990) notes how the concept of "peoples" differs from the concept of "ethnic minority" because the latter is related to ethnic migrant groups who have no ancestral sovereignty over the territory.

NOTES TO CHAPTER THREE

1. The names that I use are the most common in anthropological discourse. However, the Kogui people call themselves Kagabba and the Arhuaco people call themselves Ijku. The Wiwa people are called Arsarios, Arzarios or Malayos by western observers.
2. I will use the abbreviated name in Spanish in all cases.
3. The *Mamas* are spiritual, political leaders and ancestral authorities for the SNSM indigenous peoples.
4. It is also called the Ancient Law or Mother's Law. It refers to philosophical precepts that sustain the daily and symbolic activities of indigenous people of the SNSM.
5. "Communities" refer to settlements in several zones also referred to as "pueblos" (towns), but the term of "community" will be used here to differentiate it from the concept of *People* (Pueblos) developed in the ILO.

6. The *pagamentos* are material offerings that act as spiritual food to the spiritual mothers of species and at the same time as the essence they give so that nature may function (Villegas 1999).
7. These are the Kogui's sacred spaces (ceremonial houses) where daily aspects are interconnected to Ancient Law.
8. Indigenous peoples of the SNSM consider themselves as *elder brothers*.
9. The Lost City (Ciudad Perdida) is an archeological site in the SNSM that was discovered in 1976. It is considered one of the most important pre-Hispanic monuments on the continent for its characteristics: an entangled network of tiled roads, terraces and small circular plazas supported by walls on the steeper mountains. This City was also strategically located; therefore its inhabitants could have a variety of climates, which ranged from the tropical to the temperate to the alpine. Consequently, they had access to a great variety of agricultural resources and hunting grounds.
10. These leaders are persons who have political and academic educations and are considered new leaders because they know how to deal with the external world as well as indigenous life.
11. The Instituto Colombiano de Antropología-ICAN (now Instituto Colombiano de Antropología e Historia-ICANH) began its presence in the Ciudad Perdida in the early 1970s. ICANH is a national institution that supports anthropological, archeological and historical research that develops, defends, preserves, conserves and disseminates the national cultural patrimony and memory of the nation. Indigenous peoples and the ICANH have had discussions and conflicts for the political control of the Lost City. Indigenous peoples want to recover this sacred place.
12. The territorial strip runs from the Don Diego River in the Magdalena department to the Palomino River on the border of the Guajira Department in Colombia.
13. The Black Line refers to territorial delimitation that recognizes the ancestral territory of indigenous peoples of the SNSM, mapping the 39 most important sacred places.
14. The OGT has also begun processes of interrelation with different national institutions such as the Instituto Colombiano de Antropología e Historia-ICANH related to the political and practical control of the Lost City. However, in this dissertation, I focus only on the process of interrelation of the OGT and CTC with environmental processes and institutions as such, even though the Lost City has a strategic environmental importance for indigenous peoples of the SNSM.
15. The Learning and Innovation Loan (LIL) is a credit from the World Bank for small learning and innovation pilot projects that tend to promote replicable activities at large scale.
16. These two projects are different since the LIL is a credit from the WB to the Colombian nation for innovation and learning within the sustainable development framework that involves the national and new national-global policy executors, in this case an NGO, but that does not have certain requirements: for example, scientific analysis or explicit conservation objectives that consider local interests, a fact that makes its implementation and de-

velopment easier. Its ultimate objective is not conservation but development. The GEF Project is a donation that is dedicated exclusively to conservation which, in this case, does require involvement of the NGO with local participants.

17. This is a joint statement of the four indigenous organizations of the Sierra Nevada de Santa Marta: Wiwa Yugumaiun Bunkwanarrwa Tayrona-OWYBT Organization, Kankuamo Indigenous Organization-OIK, Indigenous Confederation Tayrona-CIT, and Gonawindúa Tayrona Organization-OGT. Valledupar, 1999.
18. Announced and explained according to the document "Policies of Indigenous People in the Sierra Nevada de Santa Marta". Cabildo Territorial Council, Organization Gonawindúa Tayrona-Indigenous Confederation Tairona-Organization Wiwa Yugumaiun Bunkwanarrwa Tayrona-Indegneous Organization Kankwama. Santa Marta 2002.
19. This recognition is granted in the national policies for protected areas and in the PDS.

NOTES TO CHAPTER FOUR

1. This idea of sacred places is one of the "new" strategies of conservation undertaken by UNESCO.
2. For information related to environmental movements see: Klandermans and Tarrow (1988), Garcîa (1992), Linkenbach (1992), Omvedt (1993), Princen (1994), Fisher (1994), Wapner (1994, 1995), Alario (1995), and Rootes (1999), among others.
3. The notion of eco-panopticon is based on the Foucault's idea of the panopticon.
4. I am grateful for the contributions of Claudia Campos (M.S. Conservation Biology) for the analysis of the environmental implications of governmental and CTC's policies.
5. Article 80 of the CPC-91.
6. Twenty of the protected areas in Colombia have been superimposed over indigenous territories, which have led to recognition of the environmental management that the indigenous exercise based on their traditional knowledge.
7. All tables in this chapter are based on the information of the PDS (FPSNSM 1997).
8. Indigenous peoples of the SNSM are demanding enlargements of their *resguardos.*
9. See Escobar (1999) for a discussion about the emergence of the idea of biodiversity.
10. Biodiversity, according to the United Nations CBD (1994) article No. 2 means "the variability among living organisms from all sources including, inter alia, terrestrial, marine and other aquatic ecosystems and the ecological complexes of which they are part; this includes diversity within species, between species and ecosystems."

11. Article 8 (j) Subject to its national legislation, respect, preserve and maintain knowledge, innovations and practices of indigenous and local communities embodying traditional lifestyles relevant for the conservation and sustainable use of biological diversity and to promote their wider application with the approval and involvement of the holders of such knowledge, innovations and practices and encourage the equitable sharing of the benefits arising from the utilization of such knowledge, innovations and practices.
12. Article 10 (c) Protect and encourage customary use of biological resources in accordance with traditional cultural practices that are compatible with conservation or sustainable use requirements.

NOTES TO CHAPTER FIVE

1. The Earth Council is "an international non-governmental organization (NGO) that was created in September 1992 to promote and advance the implementation of the Earth Summit agreements. Three fundamental objectives have guided the work of the Earth Council since its inception: to promote awareness for the needed transition to more sustainable and equitable patterns of development, to encourage public participation in decision-making processes at all levels of government, and to build bridges of understanding and cooperation between important actors of civil society and governments worldwide." (http://www.itpcentre.org/madre_index.htm)
2. The Fetzer Institute is a non-profit research and educational organization dedicated to pursuing the implications of mind-body-spirit unity in a variety of arenas.
3. I borrow and modify Tilley's term (2002) "Transnational Indigenous Peoples Movements" (TIPM) defined as a "global network of native peoples' movements and representatives—and of sympathetic institutions, non governmental organizations (NGOs) and scholars- which, through decades of international conferences, have formulated certain framing norms for indigenous politics now expressed in several international legal instruments" (Tilley 2002:526). I define TEIM as a "global network of indigenist movements, some indigenous peoples and sympathetic institutions, non-governmental organizations (NGOs) and scholars which, through decades of international conferences, have generated discourses and representations that influence global environmental policies."
4. Goldman Environmental Prizes are "awarded for sustained and important efforts to preserve the natural environment, including, but not limited to: protecting endangered ecosystems and species, combating destructive development projects, promoting sustainability, influencing environmental policies and striving for environmental justice" (http://www.goldmanprize.org/prize/about.html).
5. It is important to mention that every local, national and international meeting of indigenous peoples produces a declaration.
6. It was a result of the World Conference of Indigenous Peoples on Territory, Environment and Development, Kari-oca, Brazil, 25–30 May 1992.

7. This declaration was a result of the First International Conference on the Cultural and Intellectual Property Rights of Indigenous Peoples held in June 12–18, 1993, Aotearoa, New Zealand. In this meting over 150 delegates participated from fourteen countries, including indigenous representatives from Japan (Ainu), Australia, Cook Islands, Fiji, India, Panamá, Perú, Philippines, Surinam, USA, and Aotearoa.
8. These were indigenous peoples from Venezuela, Ecuador, Bolivia, Peru and Colombia.
9. Representatives and traditional authorities of Indigenous Peoples, Nations, and organizations from twenty eight countries, gathered from all regions of the world, including farmers, hunters, gatherers, fishers, herders, and pastoralists.
10. Lorenzo Muelas has attended international meetings to talk about his position against all the implications that genetic-resources and property-rights discourses generate for indigenous peoples.
11. Leonor Zalabata, an Arhuaco, has different articles in books and she has also been in international meetings. She is well known for her position against the Genoma project.
12. They are now very well known around the word because of their position related to yagé use and their ethical principles related to medicine.
13. COICA has been the most important indigenous peoples' organization in Latin America.
14. I convened two workshops about Indigenous Peoples and Environment with different national institutions and indigenous peoples from different cultures that were part of the methodological approach of this research.
15. Polyethnic rights are related to the legal recognition of the cultural practices of ethnic groups and religious minorities within the nation-state. Special representation rights given to ethnic groups, minorities, indigenous people and other disadvantaged or marginalized groups create a political space within the Senate for effective representation. Self-government rights are the recognition of indigenous people as "peoples" or "nations" within the state (Kymlicka 1996).

NOTES TO CHAPTER SIX

1. Spiritual leaders and representatives of the Bri-Bri, Cree, Garifuna, Keetowah, Ketchua, Kolla, Kuna, Macuna, Mapuche, Maya, Nahuat and Shuar Nations signed this declaration.
2. According to Hall (1997), three principal approaches (reflective, intentional and constructionist) have been developed to analyze representations and their relation to culture. In the reflective approach, objects, persons or ideas have meaning from the inside. That means that language operates as a mirror that reflects the meaning that lies in the "real world." In the intentional approach, social actors impose a unique meaning through language onto their world. Finally, in the constructionist approach, meaning is a social construction in which actors establish a meaningful communication with their world and the Other. These social actors construct symbolic practices and

processes in order to establish a meaningful communication. In this approach are two different perspectives: semiotic and discursive. Within the discursive perspective, Foucault is the most important proponent. This perspective is concerned "with representation as a source for the production of social knowledge—a more open system, connected in more intimate ways with social practices and questions of power" (42) and the production of knowledge through discourses within specific historical contexts and social relations.

3. The crown assigned to the Spaniards living in the cities different indigenous communities in order to establish an interchange of tools for gold or silver.
4. In 1542 indigenous peoples were considered free subjects of the King, therefore, they had to pay tribute. Paying this tribute allowed indigenous peoples to receive the "protection" of *encomenderos,* that is, to be placed under their authority.
5. *Homos silvestris* or humans with hair in all their body.
6. *Cinocéfalos* or humans with the head of a dog that lived in Brazil.
7. *Orejones* or humans with big ears that lived in California.
8. *Mestizos, mulatos and zambos* were categories used to mark peoples various racial combinations.
9. The "Environmental National Distinction" was created in 1994 by official decree No. 1125. This is a governmental distinction that recognizes persons and institutions that have dedicated part of their lives or activities to conserve and manage renewable natural resources in a sustainable way.
10. Goldman Environmental Prizes are "awarded for sustained and important efforts to preserve the natural environment, including, but not limited to: protecting endangered ecosystems and species, combating destructive development projects, promoting sustainability, influencing environmental policies and striving for environmental justice" (http://www.goldmanprize.org/prize/about.html).
11. See Bocarejo 2002 for a discussion on tradition in the SNSM.
12. The Kogui people are the first in America to receive this award. This award was conferred on Jacques Cousteau in 1996. Bio was formed by 103 countries in Greece and promotes respect for life and endorses international cooperation to protect the environment.
13. *Kayapó: Out of the Forest* (1989-52 mins), Chicago: Films Incorporated Video, Michael Beckham, directors *Blowpipes and Bulldozers (*1988-60 mins), Okley, PA: Bullfrog Films, Jeni Kennedy and Paul Tait, directors. *Fern Gully: The Last Rainforest* (1992-72 mins), 20th Century Fox Home Entertainment, Bill Kroyer, director. *Amazon Journal* (1995-58 mins), New York: Filmakers library. Geoffrey O'Connor, director. *Tong Tana: The Lost Paradise* (2001-52 mins), New York: Filmakers Library, Jan Roed, Eric Pauser, and Bjorn Cedeberg, directors. *The Shaman's Apprentice* (2001-54 mins), Okley, PA: Bullfrog Films, Miranda Smith, director (Vivanco 2002).
14. This article was written for the meeting of indigenous peoples, peasants and settlers of the SNSM within the framework of a "workshop of the regions' peoples' knowledges" in which it was noted that the main problem is the reduction of water sources.

15. Peasants in different parts of Colombia are also using the ecological identity to construct themselves and to have access to "ethnic rights" over their territory.
16. The Gaia idea is also related with notions of Mother Earth.
17. See for example the book: *Wilderness: Earth's Last Wild Places* by Russell Mittermeier, Cristina Goettsch Mittermeier, Patricio Robles Gil, Gustavo Fonseca, Thomas Brooks, John Pilgrim, and Iam R. Konstant, 2002, Cemex-CI-Agrupación Sierra Madre. The description that Conservation International gave the book states, "Conservation International's book *Hotspots* placed that organization at the forefront of global conservation efforts. *Wilderness: Earth's Last Wild Places* continues the efforts made in that previous volume, combining nearly 300 breathtaking images of untamed lands and rare glimpses of the people who inhabit them with the most current scientific analyses of their endangered ecosystems" (www.conservation.org/xp/CIWEB/publications/books_papers/books/wilderness_index.xml).
18. It is important to mention how within eco-feminism the connection between gender and the environment appears in the context of the belief that "the ideologies that legitimate injustices based in mankind, race and class are related with the ideologies that admit the exploitation and degradation of the environment" (Sturgeon 1997). According to this approach, development concepts and sustainable development can be considered in conflict with human rights, given that it is established that the environment is an important example of gender inequalities. Even though there are many positions within eco-feminism, the latter generally proposes the essentialization of the woman/nature relation and makes it an important point when considering the environment as an important aspect related to gender notions. Also, there are feminist perspectives that criticize the woman/nature relationship.
19. Turner (1995) shows how the Kayapó people passed from environmental heroes to villains when they embraced commercial activities in their territories.
20. This is one of the diverse webpages that have information about indigenous peoples and sustainable development. This is the page of the International Fund for Agricultural Development. http://www.ifad.org/events/wssd/ip/ip.htm. In Yahoo, for example, the search for indigenous peoples and sustainable development brings more than 131,000 references.
21. In June 2001 in Cartagena, Colombia, the International Association for Impact Assessment (IAIA) granted the Isoceno people a prize because of their management of natural resources.
22. You can see such products on this webpage: http://www.seacrestcrafts.com/huntinghawk/sc-arrowhead.htm
    Hunting Hawk describes her Sashes: I use virgin wool. Virgin meaning it doesn't have added wool like from old blankets, sweaters, etc. Wool sashes are best kept in a cedar box or with cedar chips to keep moths and other critters from it. Properly worn around the waist, the end of the tassle should come to about the knees. Most tribes/nations wear their sash tied on the left when at peace and on the right when at war. Though Shawnee I am adopted Cherokee Bird Clan so my finger weaving has a Cherokee influence. A Brief History of Native American Sashes and Finger Weaving: Finger

woven sashes made from inner strands of tree bark have been found in New England and dated to 3,000 years old. Initially sashes were simply used to lash hide clothing to the body. But after learning to finger weave with soft animal hair and learning what natural resources to color the hair with, the sashes became much more. With different colors, came different patterns to what finger weaving is today.

23. These are bracelets made from cotton and, according to the Kogui people, give protection to humans from bad energies and improve good energies.

NOTES TO CHAPTER SEVEN

1. Among the most important transnational NGOs supporting indigenous peoples' rights are: International Working Group for Indigenous Affairs, Cultural Survival, Survival International, World Council of Indigenous Peoples, Comission Pro-Indio, Consejo Indio de Sudamérica, Arctic Peoples' Conference, West Papuan Peoples' Front, Karen National Union, Jumma Network in Europe, Indian Council of Indigenous and Tribal Peoples, Alliance of Taiwan Aborigines, National Federation of Indigenous Peoples of the Philippines, Lumad-Mindanao, Cordillera Peoples' Alliance, Ainu Association of Hokkaido, Asia Indigenous Peoples' Pact, Naga Peoples' Movement for Human Rights, and Homeland Mission 1950 for South Moluccas, Hmong People, among others.
2. These are some of the internet pages about the Kogui, but in the international arena they are called Kogi: (http://www.entheogen.com/kogi.html; http://www.eremite.demon.co.uk/#kogi.htm; http://www.crystalinks.com/kogi.html; http://santamarta.freeservers.com/kogi.html; http://weblife.bangor.ac.uk/rem/kogi/lh_people.html; http://www.nso.lt/lucid/kogis.htm; http://www.xploreheartlinks.com/kogistory.htm; http://www.hrc.wmin.ac.uk/guest/gaia/alan.html; http://www.theosociety.org/pasadena/sunrise/42-92-3/my-rayr.htm;
3. "An important Updated Statement about the Kogui: The Kogui of Columbia, South America are under great duress and have requested cooperation on our part as well as yours. This request has come from two different Kogi organizations. Because of the war that is ensuing between the guerillas, the Colombian government and the United States, the Kogi find themselves wedged in an extremely dangerous situation. The war is all around them. Any attention on them, either in the written word or especially from people attempting to go to see them, must be stopped to save their existence. Therefore, because we at the Spirit of Ma'at truly care about the welfare of indigenous peoples of the Sierra Nevada, we are going to remove all references to them, stop posting notices and articles about them, and request that you do the same. We will not allow postings on the bulletin board either. And most of all, please do not enter Colombia attempting to physically visit them. Silence is the best gift of love we can give them at this time." With Deep Love and Respect, Drunvalo.
4. A list of Indigenous Organizations collected over the years by the Office of the High Commissioner for Human Rights (OHCHR) in Geneva is now

available on its website http://www.unhchr.ch/indigenous/indlist.htm.

5. This meeting occurred in June 1999 with the participation of more that 40 medicine men from different indigenous cultures: Ingano peoples from Caqueta, Bota Caucana, Mocoa and Valle del Sibundoy; Cofanes from valle de Guamuez, Santa Rosa de Sucumbìos, Afilador and Yarinal; Sionas from Bajo Putumayo, Kamsas del Valle de Sibundoy and the elder Carijona woman from El Tablero; Coreguajes from Orteguaza, Caqueta, and indigenous payes Tatuyos from Vaupés. Encuentro de Taitas en la Amazonía Colombiana. Unión de Médicos Indígenas Yageceros de Colombia-UMIYAC. 1999. This meeting had financial support and was organized thanks to the coalition of the Unión de Médicos Indígenas Yageceros de Colombia (UMIYAC), the Ingana organization Tanda Chiridu Inganokuna and the international environmental NGO, Amazon Conservation Team.
6. Sponsored by: Organización Pies Descalzos, PNUD-United Nations program for the development, Pro SNSM Foundation, Presidency of the Republic, Ministry of Environment, Ministry of Cultura, Ministry of Education, OGT, Indigenous Tayrona Zhigumanin Confederation, Bunquannarua Tayrona, Owybt, Indupalma and Tourist Promotion Fund of Colombia.
7. Tebtebba means, "*discourse*" in the Philippine indigenous *Kankanaey* dialect.
8. The meeting was lead by Rigoberta Menchu Tum, Nobel Peace Prize and Good Will Ambassador of the International Year of Indigenous Peoples, representing the Secretary General of the United Nations.
9. I appreciate the contribution from the Biologist, Claudia Campos, in the biological-implication analysis of the governmental and CTC's policies.
10. Such as World Wildlife Fund, International Union for Conservation of Nature-IUCN, and Conservation International-CI, among others.
11. For indigenous peoples of the SNSM this is the term that represents the interrelation between the spiritual and the material that generates the vital principle of life.
12. Indigenous peoples' environmental practices are defined by many anthropologists and biologists as less exploitative and in harmony with the ecosystems, see for example, *Indigenous Traditions and Ecology.* John A. Grim (ed.). Cambridge: Harvard University, Center for the Study of Old Religions (2001).

References

Achito, Alberto. "Los pueblos indígenas y medio ambiente, una propuesta de paz." In *Ambiente Para la Paz*. Bogotá: Ministerio del Medio Ambiente, Cormagdalena, 1998.

Alario, Margarita. *Environmental Destruction, Risk Exposure and Social Asymmetry. Case of the Environmental Movements' Actions*. London: University Press of America, 1995.

Albo, Xavier. "Our Identity Starting from Pluralism in the Base." In *The Postmodernism Debate in Latin America,* edited by John Beverley, Michael Aronna, and José Oviedo. London: Duke University Press, 1995.

Alcorn, Janis B. "Noble Savage or Noble State?: Northern Myths and Southern Realities in Biodiversity Conservation." *Etnoecológica* 2, no. 3 (1994).

Alonso, Germán. "Biodiversidad y derechos colectivos de los pueblos indígenas y locales en Colombia." In *Ambiente para la paz*. Bogotá: Ministerio del Medio Ambiente, Cormagdalena, 1998.

Álvarez, Jairo Hernán. "Movimiento ambiental colombiano." In *Se hace el camino al andar. Aportes para una historia del movimiento ambiental en Colombia.* Bogotá: Ecofondo, 1997.

Álvarez, Sonia, Evelina Dagnino, and Arturo Escobar, ed. *Cultures of Politics, Politics of Cultures. Re-visioning Latin America Social Movements.* Boulder: Westview Press, 1998.

———. "Introduction." In *Cultures of Politics, Politics of Cultures. Re-visioning Latin America Social Movements,* edited by Sonia Alvarez, Evelina Dagnino and Arturo Escobar. Boulder: Westview Press, 1998(a).

Anderson, Terry L., and Donald R. Leal. *Free Market Environmentalism.* Boulder: Westview Press, 1991.

Archila, Mauricio, and Mauricio Pardo, ed. *Movimientos sociales, Estado y democracia en Colombia.* Bogotá: ICANH, CES, Universidad Nacional de Colombia, 2001.

Arnold, David. *La naturaleza como problema histórico. El medio, la cultura y la expansión de Europa.* México: Fondo de Cultura Económica, 2000.

Arvelo-Jiménez, Nelly. "Los pueblos indígenas y las tesis ambientalista sobre el manejo global de sus recursos." In *Articulación de la diversidad. Pluralidad étnica, autonomía y democratización en América Latina, Grupo de Barbados,* comp.: Georg Grunberg. Quito: Biblioteca Abya-Yala 27, 1995.

Assies, William, Gemma Van der Haar, and André Hoekema, ed. *The Challenge of Diversity. Indigenous Peoples and Reform of the State in Latin America.* Amsterdam: Thela Thesis, 1998.

Avirama, Jesús, and Rayda Márquez. "The Indigenous Movement in Colombia." In *Indigenous Peoples and Democracy in Latin America,* edited by Donna Lee Van Cott. New York: St. Martin's Press, 1994.

Baptiste, Luis Guillermo, and Sarah Hernández. "Elementos para la valoración económica de la biodiversidad colombiana." In *Diversidad biológica y cultural,* edited by Margarita Flórez. Bogotá: Ilsa, 1998.

Barth, Frederick. *Ethnic Groups and Boundaries. The Social Organization of Cultural Difference.* London: George Allen & Unwin, 1969.

Bartra, Roger. *The Cage of Melancholy.* New Brunswick, New Jersey: Rutgers University Press, 1992.

Bebbington, Anthony, et al. *Actores de una década ganada. Tribus, comunidades y campesinos en la modernidad.* Quito: Comunidec, 1992.

Behar, Ruth, Beborah Gordon, ed. *Women Writing Culture.* Berkeley: University of California Press, 1995.

Beneria-Surkin, J. "De guerreros a negociadores: un análisis de la sustentabilidad de la Capitanía del Alto y Bajo Izozog -CABI-. Estrategias de conservación y desarrollo." In *Desarrollo sostenible en la Amazonía. ¿Mito o realidad?,* edited by Mario Hiraoka, and Santiago Mora. Quito: Ediciones Abya-Yala, Colección Hombre y Ambiente, 63–64, número monográfico, 2001.

Bengoa, José. *La emergencia indígena en América Latina.* Santiago de Chile: Fondo de Cultura Económica, 2000.

Berraondo, Miguel. *Los derechos medioambientales de los pueblos indígenas. La situación de la región amazónica.* Quito: Ediciones Abya-Yala, 1999.

Bhabha, Homi. *The Location of Culture.* London: Routledge, 1994.

Blatter, Joachim, Helen Ingram, and Pamela Doughman. "Emerging Approaches to Comprehend Changing Global Contexts." In *Reflections on Water: New Approaches to Transboundary Conflicts and Cooperation,* edited by Joachim Blatter, and Helen Ingram. London: The MIT Press, 2001.

Blatter, Joachim, Helen Ingram, Helen, and Suzanne Levesque. "Expanding Perspectives on Transboundary Water." In *Reflections on Water: New Approaches to Transboundary Conflicts and Cooperation,* edited by Joachim Blatter, and Helen Ingram. London: The MIT Press, 2001.

Bocarejo, Diana. *Indigenizando lo 'Blanco': Conversaciones con Arhuacos y Koguis de la Sierra Nevada de Santa Marta.* Bogotá: Manuscript, 2002.

Bonfil Batalla, Guillermo. *Utopía y revolución: el pensamiento político contemporáneo de los indios en América Latina.* México: Editorial Nueva Imagen, 1981.

Bourdieu, Pierre. *Outline of a Theory of Practice.* Cambridge: Cambridge University Press, 1977.

———. *The Field of Cultural Production: Essays on Art and Literature.* Cambridge: Polity Press, 1993.

Borja, Jaime. *Del bárbaro y de la naturaleza agreste. Una historia moral del indio neogranadino.* Bogotá: Manuscript, 2002.

Brading, David A. "La historia natural y la civilización amerindia."In *Descubrimiento, conquista y colonización de América a quinientos años,* comp.: Carmen Bernand. México: Consejo Nacional para la Cultura y las Artes, Fondo de Cultura Económica, 1994.

Brosius, J. Peter. "Green Dots, Pink Hearts: Displacing Politics from the Malaysian Rain Forest." *American Anthropologist* 101, no. 1 (1999).

———. "Analyses and Interventions. Anthropological Engagements with Environmentalism." *Current Anthropology* 40, no. 3 (1999a).

———. "Endangered Forest, Endangered People: Environmentalist Representations of Indigenous Knowledge." In *Indigenous Environmental Knowledge and its Transformations. Critical Anthropological Perspectives,* edited by Roy Ellen, Peter Parkes, and Alan Bicker. Amsterdam: Overseas Publisher Association, 2000.

Brosius, J. Peter, Anne Tsing, and Charles Zerner. "Representing Communities: Histories and Politics of Community-based Resource Management." *Society and Natural Resources* 11, no. 2 (1998).

Brush, Stephen, and Doreen Stabinsky, ed. *Valuing Local Knowledge. Indigenous People and Intellectual Property Rights.* Washington: Island Press, 1996.

Bryant, Raymond, and Sinead Bailey. *Third World Political Ecology.* London: Routledge, 1997.

Brysk, Alison. "Social Movements, the International System, and Human Rights in Argentina." *Comparative Political Studies* 26, no. 3 (1993).

———. "Acting Globally: Indian Rights and International Politics in Latin America." In *Indigenous Peoples and Democracy in Latin America,* edited by Donna Lee Van Cott. New York: St. Martin's Press, 1994.

———. "Turning Weakness into Strength: The Internationalization of Indians' Rights. "*Latin American Perspectives* (Special Issue on Ethnicity and Class in Latin America) 23, no. 2 (1996).

———. *From Local Village to Global Village: Indian Rights and International Relations in Latin America.* Stanford: Stanford University Press, 2000.

Cabarcas, Hernando. *Bestiario del Nuevo Reino de Granada.* Bogotá: Colcultura, 1994.

Caldeira, Teresa. *City of the Walls: Crime, Segregation, and Citizenship in Sâo Paulo.* Berkeley: University of California Press, 2001.

Calderón, Fernando. "Latin America Identity and Mixed Temporalities." In *The Postmodernism Debate in Latin America,* edited by John Beverley, Michael Aronna, and José Oviedo. London: Duke University Press, 1995.

Calderón, Fernando, Alejandro Piscitelli, and José Luis Reyna. "Social Movements: Actors, Theories, Expectations." In *The Making of Social Movements in Latin America,* edited by Arturo Escobar, and Sonia Alvarez. Boulder: Westview Press, 1992.

Campos, Claudia. *El etnoturismo y el ecoturismo en territorios de pueblos indígenas en Colombia.* Bogotá: Manuscript, 2002.

Carrizosa Umaña, Julio. "Algunas raíces del ambientalismo en Colombia: estética, nacionalismo y prospectiva." In *Se hace el camino al andar. Aportes para una historia del movimiento ambiental en Colombia.* Bogotá: Ecofondo, 1997.

Castells, Manuel. *The Power of Identity.* Malden, Oxford: Blackwell, 1997.

Cayón, Luis. *Manejo ecológico de los indígenas de la Sierra Nevada de Santa Marta.* Bogotá: Manuscript, 2002.

Chaves, Margarita. "Discursos subalternos de identidad y movimiento indígena en el Putumayo." In *Movimientos sociales, Estado y democracia en Colombia,* edited by Mauricio Archila, and Mauricio Pardo. Bogotá: ICANH, CES, Universidad Nacional de Colombia, 2001.

Clavero, Bartolomé. "Pronunciamientos indígenas de las constituciones americanas." http://alertanet.org/constitucion-indigenas.htm, 2003.

Clifford, James. *The Predicament of Culture: Twentieth-century Ethnography, Literature, and Art.* Cambridge: Harvard University Press, 1988.

Clifford, James, and George Marcus, ed. *Writing Culture: The Poetics and Politics of Ethnography.* Berkeley: University of California Press, 1986.

Cohen, Jean. "Strategy or Identity. New Theoretical Paradigms and Contemporary Social Movements." *Social Research* 52, no. 4 (1985).

COICA (Coordinadora de las Organizaciones Indígenas de la Cuenca Amazónica). *Los pueblos indígenas amazónicos y su participación en la agenda internacional.* Bogotá: COICA, 2001.

Colchester, Marcus. "Indigenous Rights and the Collective Conscious." *Anthropology Today* 18, no. 1 (2002).

Collier, George. "Aboriginal Sin and the Garden of Eden: Humanist Views of the Amerindian", Stanford: Department of Anthropology, Stanford University, n.d.

Collier, Jane, Bill Maurer, and Liliana Suárez Navas. "Sanctionated Identities: Legal Construction of Modern Personhood." *Identities* 2, no. 1–2 (1995).

Collinson, Hellen, ed. *Green Guerrillas. Environmental Conflicts and Initiatives in Latin America and the Caribbean.* London: Latin America Bureau, 1996.

Comaroff, John L., and Jean Comaroff. *Of Revelation and Revolution.* Chicago: University of Chicago Press, vol. 2, 1997.

Conklin, Beth. "Body Paint, Feathers, and Vcrs: Aesthetics and Authenticity in Amazonian Activism:" *American Ethnologist* 24, no. 4 (1997).

———. "Shamans versus Pirates in the Amazonian Treasure Chest." *American Anthropologist* 104, no. 4 (2002).

Conklin, Beth, and Laura Graham. "The Shifting Middle Ground: Amazonian Indians and Eco-Politics." *American Anthropologist* 97, no. 4 (1995).

CPC *(Constitución Política de Colombia),* Bogotá: Ediciones Impreandes, 1991.

Coombe, Rosemary. *The (c)ultural Life of Intellectual Properties.* Durham: Duke University Press, 1998.

Correa, François, ed. *La selva humanizada.* Bogotá: Ican, Cerec, 1990.

Cronon, William. *Uncommon Ground, Rethinking Human Place in Nature.* New York: W. W. Norton & Company, 1995.

Crosby Jr., Alfred W. *El imperialismo ecológico. La expansión biológica de Europa.* Barcelona: Editorial Crítica, 1988.

CTC (Consejo Territorial de Cabildos). "Declaración conjunta de las cuatro organizaciones indígenas de la Sierra Nevada de Santa Marta para la interlocución con el Estado y la sociedad nacional." Valledupar: 1999.

———. "Políticas de los pueblos indígenas de la Sierra Nevada de Santa Marta."Santa Marta: 2000.

———. "Declaración del Consejo Territorial de Cabildos ante el Consejo Ambiental Regional de la Sierra Nevada de Santa Marta." Santa Marta: 2001.

———. "Políticas de los pueblos indígenas de la Sierra Nevada de Santa Marta." Santa Marta: 2001(a).

———. "Políticas de los pueblos indígenas de la Sierra Nevada de Santa Marta." Santa Marta: 2002.
———. "Propuesta del Consejo Territorial de Cabildos ante el comité directivo del PDS." Santa Marta: 2002(a).
Dagnino, Evelina. "Culture, Citizenship, and Democracy: Changing Discourses and Practices of the Latin America Left." In *Cultures of Politics, Politics of Cultures. Re-visioning Latin America Social Movements,* edited by Sonia Alvarez, Evelina Dagnino, and Arturo Escobar. Boulder: Westview Press, 1998.
Dalton, Rex. "Tribes Query Motives of Knowledge Databases." *Nature* 419, no. 866 (2002).
Danier, Eric, ed. *Discourses of the Environment.* Oxford: Blackwell Publishers, 1999.
Descola, Philippe. "Constructing Natures: Symbolic Ecology and Social Practice." In *Nature and Society. Anthropological Perspectives,* edited by Philippe Descola, and Gisli Pálsson. London: Routledge, 1996.
Descola, Philippe, and Gisli Pálsson, ed. *Nature and Society. Anthropological Perspectives.* London: Routledge, 1996.
Díaz Polanco, Héctor. *Autonomía regional: la autodeterminación de los pueblos indios.* México: Siglo XXI Editores, 1991.
Dobson, Andrew. *Green Political Thought.* London: Routledge, 1989.
Dover, Robert V. H., Joanne Rappaport. "Introduction." *Journal of Latin America Anthropology* 1, no. 2 (1996).
Dowie, Mark. *Losing Ground: American Environmentalism at the Close of the 20th Century.* Cambridge: MIT Press, 1995.
Edelman, Marc. "Social Movements: Changing Paradigms and Forms of Politics." *Annual Review of Anthropology* 30 (2001).
Ellen, Roy. "Introduction." In *Redefining Nature. Ecology, Culture and Domestication,* edited by Roy Ellen, and Katsuyoshi Fukui. Oxford: Berg, 1996(a).
———. "The Cognitive Geometry of Nature: a Contextual Approach." In *Nature and Society. Anthropological Perspectives,* edited by Philippe Descola, and Gisli Pálsson. London: Routledge, 1996(b).
Ellen, Roy, and Katsuyoshi Fukui, ed. *Redefining Nature. Ecology, Culture and Domestication.* Oxford: Berg, 1996.
Ellen, Roy, Peter Parkes, and Alan Bicker, ed. *Indigenous Environmental Knowledge and its Transformations. Critical Anthropological Perspectives.* Amsterdam: Overseas Publisher Association, 2000.
Epstein, Barbara. "Rethinking Social Movements Theory." *Socialist Review 20,* no. 1 (1990).
Escobar, Arturo. "Culture, Economics, and Politics in Latin American Social Movements Theory and Research." In *The Making of Social Movements in Latin America,* Boulder, edited by Arturo Escobar, and Sonia Alvarez. Boulder: Westview Press, 1992.
———. *Encountering Development: The Making and Unmaking of the Third World.* Princeton: Princeton University Press, 1995.
———. "Constructing Nature. Elements for a Poststructural Political Ecology." In *Liberation Ecologies. Environment, Development and Social Movements,* edited by Richard Peet, and Michael Watts. London: Routledge, 1996.
———. "Whose Knowledge, Whose Nature? Biodiversity, Conservation, and the Po-

litical Ecology of Social Movements." *Journal of Political Ecology* 5 (1998).
———. *El final del salvaje. Naturaleza, cultura y política en la antropología contemporánea.* Bogotá: ICANH, CEREC, 1999.

Escobar, Arturo, and Sonia Alvarez, ed. *The Making of Social Movements in Latin America.* Boulder: Westview Press, 1992.

———. "Introduction: Theory and Protest in Latin America Today." In *The Making of Social Movements in Latin America,* Boulder, edited by Arturo Escobar, and Sonia Alvarez. Boulder: Westview Press, 1992 (a).

Escobar, Arturo, and Álvaro Pedrosa, ed. *Pacífico, ¿desarrollo o diversidad?* Bogotá: CEREC, Ecofondo, 1996.

Fairhead, James and Melissa Leach. *Misreading the African Landscape. Society and Ecology in a Forest-savanna* Mosaic. Cambridge: Cambridge University Press. 1995.

Fals Borda, Orlando, and C. Rodríguez. *Investigación participativa.* Montevideo: Ediciones de la Banda Oriental, 1987.

Ferguson, Jim. *The Anti-Politics Machine: Development, Depolitization, and Bureaucratic Power in Leshoto.* Cambridge, New York: Cambridge University Press, 1990.

Findji, María Teresa. "The Indigenous Authorities Movement in Colombia. " In *The Making of Social Movements in Latin America,* Boulder, edited by Arturo Escobar, and Sonia Alvarez. Boulder: Westview Press, 1992.

Fisher, William. "Megadevelopment, Environmentalism, and Resistance: The Institutional Context of Kayapo Indigenous Politics in Central Brazil." *Human Organization.* 53, no. 3 (1994).

Flórez, Margarita. *Protección del conocimiento tradicional y tratamiento legal internacional de los pueblos indígenas.* Bogotá: Manuscript, 2001.

Foucault, Michel. "Questions of Method." In *The Foucault Effect: Studies in Governmentality,* edited by Graham Burchell, Colin Gordon, and Peter Miller. Chicago: University Chicago Press, 1991.

———. "Governmentality." In *The Foucault Effect: Studies in Governmentality,* edited by Graham Burchell, Colin Gordon, and Peter Miller. Chicago: University Chicago Press, 1991a.

FPSNSM (Fundación Pro Sierra Nevada de Santa Marta).PDS *Plan de desarrollo sostenible de la Sierra Nevada de Santa Marta. Estrategias de conservación de la Sierra Nevada de Santa Marta.* Santa Marta: Proyecto de Cooperación Colombo-Alemán, MMA, DNP, Consejería Presidencial para la Costa Atlántica, Gobernación del Magdalena, Gobernación del Cesar, Gobernación de la Guajira, Carbocol y Embajada Real de los Países Bajos, 1997.

———. *Evaluación ecológica rápida. Definición de áreas críticas para la conservación en la Sierra Nevada de Santa Marta.* Santa Marta: FPSNSM, 1998.

García, María. "The Venezuelan Ecology Movement." In *The Making of Social Movements in Latin America,* Boulder, edited by Arturo Escobar, and Sonia Alvarez. Boulder: Westview Press, 1992.

García Canclini, Néstor. "The Hybrid: A Conversation with Margarita Zires, Raymundo Mier, and Mabel Piccini." In *The Postmodernism Debate in Latin America,* edited by John Beverley, Michael Aronna, and José Oviedo. London: Duke University Press, 1995.

Ginsburg, Faye. "Embedded Aesthetics: Creating a Discursive Space for Indigenous Media." *Cultural Anthropology* 9, no. 3 (1994).

González, Pablo. "Some Reflections on Liberation Struggles in Latin America." In *New Social Movements in the South,* edited by Ponna Wignaraja. London, New Jersey: Zed Books, 1993.

Gragson, Ted, and Ben Blount, ed. *Ethnoecology. Knowledge, Resources and Rights.* Atenas, London: University of Georgia Press, 1999.

Grim, John (ed.). *Indigenous Traditions and Ecology.* Cambridge, Harvard University, Center for the Study of Old Religions, 2001.

Griswold, Esther. "State Hegemony at Large: International Law and Indigenous Rights." *Polar* 19, no. 1 (1995).

Gros, Christian. *Colombia Indígena. Identidad cultural y cambio social.* Bogotá: CEREC, 1991.

———. "Indigenismo y etnicidad: el desafío neoliberal." In *Antropología en la modernidad,* edited by María Victoria Uribe, and Eduardo Restrepo. Bogotá: ICAN, 1997.

———. *Ser diferente por (para) ser moderno, o la paradoja de la modernidad.* Bogotá: Manuscript, 1998.

———. *Políticas de la etnicidad: identidad, Estado y modernidad.* Bogotá: ICANH, 2000.

Grove, Richard. *Green Imperialism.* Cambridge: Cambridge University Press, 1995.

Grueso, Libia, Carlos Rosero, and Arturo Escobar. "The Process of Black Community Organization in the Southern Pacific Coast Region of Colombia." In *Cultures of Politics, Politics of Cultures. Re-visioning Latin America Social Movements,* edited by Alvarez, Sonia, Evelina Dagnino and Arturo Escobar. Boulder: Westview Press, 1998.

Grunberg, Georg. "Los indios de la selva: nuevas estrategias y nuevas alianzas." In *Articulación de la diversidad. Pluralidad étnica, autonomía y democratización en América Latina, Grupo de Barbados,* comp.: Georg Grunberg. Quito: Biblioteca Abya-Yala 27, 1995.

Guha, Ramachandra. *Environmentalism. A Global History.* New York: Logman, 2000.

Gupta, Akhil. *Postcolonial Developments. Agriculture in the Making of Modern India.* Durham: Duke University Press, 1998.

Gupta, Akhil, and James Ferguson, eds., *Anthropological Locations.* Berkeley: University of California Press, 1997.

Gutiérrez, Óscar. "Biodiversity and New Drugs." *Trends in Pharmacological Sciences* 23, no.1, 2002.

Haas, Peter. "Do Regimes Matter? Epistemic Communities and Mediterranean Pollution Control." *International Organisation* 43, no. 3 (1989).

———. "Introduction: Epistemic Communities and International Policy Coordination." *International Organisation* 46, no. 1 (1992).

Hale, Charles. "Does Multiculturalism Menace? Governance, Cultural Rights and the Politics of Identity in Guatemala." *J. Latin Am. Stud* 34 (2002).

Hall, Stuart. "Cultural Identity and Diaspora." In *Identity: Community, Culture, Difference,* edited by Jonathan Rutherford. London: Lawrence & Wishart, 1990.

———. *Representation. Cultural Representations and Signifying Practices*. London: Sage Publications, 1997.

Haraway, Donna. *Simians, Cyborgs and Women: The Reinvention of Nature*. New York: Routledge, 1991.

———. *Modest_Witness@Second_Millennium. FemaleMan(c)_Meets_Oncomouse (TM). Feminism and Technoscience*. London: Routledge, 1997.

Hellman, Judith Adler. "The Study of New Social Movements in Latin America." In *The Making of Social Movements in Latin America,* Boulder, edited by Arturo Escobar, and Sonia Alvarez. Boulder: Westview Press, 1992.

Hernández, Aída, and Francisco Ortiz. "Las demandas de la mujer indígena en Chiapas." *Nueva Antropología* 49 (1994).

Hildebrand, Martin von. "Cosmovisión y el concepto de enfermedad entre los ufaina." In *Medicina, shamanismo y botánica,* edited by Myriam Jimeno, and Adolfo Triana. Bogotá: FUNCOL, 1983.

Hodgson, Dorothy. "Comparative Perspectives on the Indigenous Rights Movements in Africa and the Americas." *American Anthropologist* 104, no. 4 (2002).

Holston James and Teresa Caldeira. "Democracy, law, and violence: disjunctions of Brazilian citizenship." In *Fault Lines of Democracy in Post-Transition Latin America*, edited by Felipe Aguero, and Jeffrey Stark. Miami: University of Miami North-South Center Press, 1998.

ICANH, Instituto von Humboldt, Colciencias. *Memorias del Taller Nacional Pueblos Indígenas, Comunidades Tradicionales y Medio Ambiente*. Bogotá: Manuscript, 2002.

Ingold, Tim. "Hunting and Gathering as Ways of Perceiving the Environment." In *Redefining Nature. Ecology, Culture and Domestication,* edited by Roy Ellen, and Katsuyoshi Fukui. Oxford: Berg, 1996a.

Ingold, Tim. "The Optimal Forager and Economic Man." In *Nature and Society:Anthropological Perspective,* edited by Philippe Descola, and Gisli Pálsson. London: Routledge, 1996b.

Jackson, Jean. "Being and Becoming an Indian in the Vaupés." In *Nation-States and Indians in Latin America,* edited by Greg Urban, and Joel Scherzer. Austin: University of Texas Press, 1991.

———. "The Impact of the Recent National Legislation in the Vaupés Region of Colombia." *Journal of Latin America Anthropology* 1, no.2 (1996).

Kauffman, L. A. "The Antipolitics of Identity." *Socialist Review* 20, no.1 (1990).

Klandermans, Bert, and Sidney Tarrow. "Mobilization into Social Movements: Synthesizing European and American Approaches." *International Social Movement Research: A Research Annual* 1 (1988).

König, Hans-Joachim et al. *El indio como sujeto y objeto de la historia latinoamericana. Pasado y presente.* Frankfurt: Vervuert, American Eystettensin, 1998.

Kymlicka, Will. "Three Forms of Group Differentiated Citizenship in Canada." In *Democracy and Difference: Contesting the Boundaries of the Political,* edited by Seyla Benhabib. Princeton: Princeton University Press, 1996.

Latour, Bruno. *We Have Never Been Modern.* Cambridge, Massachusetts: Harvard University Press, 1993.

Laurent, Virginie. "Pueblos indígenas y espacios políticos en Colombia." In *Modernidad, identidad y desarrollo,* edited by María Lucía Sotomayor. Bogotá: Ican, 1998.

———. *Pueblos indígenas y espacios políticos en Colombia. Motivaciones, campos de acción e impactos (1990–1998), informe final, convocatoria Becas Nacionales Ministerio de Cultura.* Bogotá: Manuscript, 2001.

Lazarus-Black, Mindie, and Susan F. Hirsch. *Contested States. Law, Hegemony and Resistance.* London: Routhedgc, 1994.

Leff, Enrique. *La geopolítica de la biodiversidad y el desarrollo sustentable: economización del mundo, racionalidad ambiental y reapropiación social de la naturaleza.* Bogotá: Manuscript, 2002.

Linkenbach, Antje. "Ecological Movements and the Critique of Development: Agents and Interpreters." *Thesis Eleven* 39 (1992).

Little, Paul E. "Environments and Environmentalism in Anthropological Research: Facing a New Millennium." *Annual Review of Anthropology* 28 (1999).

Long Martello, Marybeth. "A Paradox of Virtue?: 'Other' Knowledges and Environment-Development Politics." *Global Environmental Politics* 1, no. 3 (2001).

Luca, Frank. *Serpents in the Garden and the Amerindian 'Paradise Lost' or Rattlesnakes and Indians, Real and Imaginary: An Examination of the 'Ecological Indian' Myth, 1492–1992.* Bogotá: Manuscript, 2001.

Luke, Timothy. *Ecocritique: Contesting the Politics of Nature, Economy, and Culture.* Minneapolis-London: University of Minnesota Press, 1997.

———. *Capitalism, Democracy, and Ecology.* Urbana, Chicago: University of Illinois Press, 1999.

———. "Environmentality as Green Governmentality." In *Discourses of the Environment, edited by* Éric Danier. Oxford: Blackwell Publishers, 1999(a).

Lutz, Catherine, and Jane Collins. *Reading National Geographic.* Chicago, London: University of Chicago Press, 1993.

McAdams, Doug, J. McCarthy, and Mayer Zald. "Opportunities, Mobilizing Structures, and Framing Process—Toward a Synthetic, Comparative Perspective on Social Movements." In *Comparative Perspectives on Social Movements: Political Opportunities, Mobilizing Structures, and Cultural Framings,* edited by Doug McAdams, J. McCarthy, and Mayer Zald. Cambridge: Cambridge University Press, 1996.

McAdams, Doug, Sidney Tarrow, and Charles Tilly. "Towards an Integrated Perspective on Social Movements and Revolution." In *Ideals, Interests, and Institutions: Advancing Theory in Comparative Politics,* edited by Marc Irving Lichbach, and Alan Kuckerman. Cambridge: Cambridge University Press, 1997.

McAfee, Kathleen. "Selling Nature to Save it? Biodiversity and Green Developmentalism." *Society and Space* 17, no. 2 (1999).

Mallon, Florencia. "Indian Communities, Political Cultures, and the State in Latin America, 1780-1990." *Journal of Latin American Studies* 24, Quincentenary Supplement (1992).

Marcus, George. "Ethnography in/of the World System: The Emergence of Multisited Ethnography." *Annual Review of Anthropology* 24 (1995).

Marcus, George, and Michael Fischer. *Anthropology as a Cultural Critique.* Chicago: University of Chicago Press, 1986.

Márquez, Germán. "Notas para una historia de la ecología y su relación con el movimiento ambiental en Colombia." In *Se hace el camino al andar. Aportes*

para una historia del movimiento ambiental en Colombia. Bogotá: Ecofondo, 1997.
Maurer, Bill. *Recharting the Caribbean: Land, Law, and Citizenship in the British Virgin Islands*. Ann Arbor: University Michigan Press, 2000.
Maxwell, Anne. *Colonial Photography and Exhibitions. Representations of the 'Native' and the Making of European Identities*. London, New York: Leicester University Press, 1999.
Melo, Jorge Orlando. "Algunas consideraciones globales sobre 'modernidad' y 'modernización'." In *Colombia al despertar de la modernidad,* edited by Fernando Viviescas, and Fabio Giraldo Isaza. Bogotá: Foro Nacional por Colombia, 1998.
Melucci, Alberto. "The New Social Movements: A Theoretical Approach." *Social Science Information. Information sur les Sciences Sociales*. London: Sage Publications, 1980.
———. "The Symbolic Challenge of Contemporary Movements." *Social Research* 52, no. 4 (1985).
———. "Social Movements and the Democratization of Everyday Life." In *Civil Society and the State,* edited by John Keane. New York: Verso, 1993.
Merry, Sally Engle. "Crossing Boundaries: Ethnography in the Twenty-First Century." *Polar* 23, no.2 (2000).
Milton, Kay. *Environmentalism and Cultural Theory.* London, New York: Routledge, 1996.
Mittermeier, Russell, et al. *Wilderness: Earth's Last Wild Places*. México: Cemex-CI-Agrupación Sierra Madre, 2002.
MMA. *Proyecto colectivo ambiental.* Bogotá: MMA, 2000.
Moore, Donald. "Contesting Terrain in Zimbabwe's Eastern Highlands: Political Ecology, Ethnography, and Peasant Resources Struggles." *Economic Geography* 69, no. 4 (1993).
———. "Marxism, Culture, and Political Ecology. Environmental Struggles in Zimbabwe's Eastern Highlands." In *Liberation Ecologies. Environment, Development and Social Movements,* edited by Richard Peet, and Michael Watts. London: Routledge, 1996.
———. "Clear Waters and Muddied Histories: Environmental History and the Politics of Community in Zimbabwe's Eastern Highlands." Manuscript, 1997.
Mora, Santiago. "La metáfora ecológica: simbolismo y conservación." *Revista Colombiana de Antropología* XXXII (1995).
Moseley, K. P. "In Defense of the Primitive." In *Rethinking the Third World,* edited by Rosemary E. Gali. New York: Taylor & Francis, 1991.
Motzafi-Haller, Pnina. "Liberal Discourses of Cultural Diversity and Hegemonic Constructions of Difference: Basarwa in Contemporary Botswana." *Polar* 2 (1995).
Muelas, Lorenzo. "Acceso a los recurso de la biodiversidad y pueblos indígenas." In *Diversidad biológica y cultural,* edited by Margarita Flórez. Bogotá: Ilsa, 1998.
Nash, June. *Mayan Visions. The Quest for Autonomy in an Age of Globalization.* London: Routledge, 2001.
Navarro Hernández, Antonio. *El proceso histórico de ocupación del área de Santa Marta: elementos para el análisis del ordenamiento de su territorio rural.* Santa Marta: Manuscript, 2001.

Nazarea, Virginia D, ed. *Ethnoecology, Situated Knowledge/Located Lives.* Tucson: University of Arizona Press, 1999.

Nelson, Diane. *A Finger in the Wound. Body Politics in Quincentennial Guatemala.* Los Angeles: University of California Press, 1999.

Nieto Olarte, Mauricio. *Remedios para el imperio. Historia natural y la apropiación del Nuevo Mundo.* Bogotá: ICANH, 2000.

Nochlin, Linda. *The Politics of the Vision.* New York: Harper & Row Publishers, 1989.

Offe, Claus. "New Social Movements: Challenging the Boundaries of Institutional Politics." *Social Research* 52, no.4 (1985).

OGT (Organización Gonawindúa Tayrona). *La ilusión de Serankua.* Santa Marta: Manuscript, 1998.

OGT (Organización Gonawindúa Tayrona), UAESPNN (Unidad Administrativa Especial del Sistema de Parques Nacionales Naturales), DGAI (Dirección General de Asuntos Indígenas del Ministerio del Interior*). Lineamientos generales de manejo intercultural del Parque Sierra Nevada de Santa Marta. Documento de trabajo para concertar con las cuatro organizaciones indígenas de la Sierra Nevada de Santa Marta.* Santa Marta: Manuscript, 2000.

Omvedt, Gail. *Reinventing Revolution: New Social Movements and the Social Tradition in India.* Armonk: M. E. Sharpe, 1993.

ONIC (Organización Nacional Indígena). "La paz y los pueblos indígenas." *Ambiente para la paz.* Bogotá: MMA, Cormagdalena, 1998.

Oquist, Paul. *Violence, Conflict, and Politics in Colombia.* London: Academic Press, 1980.

Orewa (Organización Regional Indígena Embera-Wounan*). Lo que queremos y pensamos hacer en nuestro territorio.* Quibdó: Manuscript, 1996.

Ortiz, Héctor, and Rosalba Aída Hernández. "Constitutional Amendments and New Imaginings of the Nation: Legal Anthropology and Gendered Perspectives on Multicultural Mexico." *Polar* 19, no.1 (1996).

Orrantia, Juan Carlos. *Matices kogui y entidades discontinuas. Representaciones y negociación en la marginalidad.* Bogotá: Manuscript, 2002.

Padilla, Guillermo. "Derecho mayor y derecho constitucional; comentarios en torno a sus confluencias y conflictos." *Pueblos indios, soberanía y globalismo,* comp.: Stefano Varese. Quito: Ediciones Abya-Yala, no.32, 1996.

Palacio, Germán. "Caminando con el paso al frente." In *Se hace el camino al andar. Aportes para una historia del movimiento ambiental en Colombia.* Bogotá: Ecofondo, 1997.

Pálsson, Gisli. "Human-Environmental Relations: Orientalism, Paternalism and Communism." In *Nature and Society. Anthropological Perspectives,* edited by Philippe Descola, and Gisli Pálsson. London: Routledge, 1996.

Pardo, Mauricio. "Movimientos sociales y actores no gubernamentales." In *Antropología en la modernidad,* edited by María Victoria Uribe and Eduardo Restrepo. Bogotá: ICAN, 1997.

Peet, Richard; Watts, Michael. *Liberation Ecologies. Environment, Development and Social Movements.* London: Routledge, 1996.

Plotke, David. "What's so New About New Social Movements?." *Socialist Review* 20, no.1 (1990).

Princen, Thomas. "NGOs: Creating a Niche in Environmental Diplomacy." In *Envi-*

ronmental NGOs in World Politics. Linking the Local and the Global, edited by Thomas Princen, and Matthias Finger. London, New York: Routledge, 1994.

Quijano, Aníbal. "Modernity, Identity, and Utopia in Latin America." In *The Postmodernism Debate in Latin America,* edited by John Beverley, Michael Aronna, and José Oviedo. London: Duke University Press, 1995.

Quiñones, Omar Ernesto. "Territorios étnicos, ambiente y paz." *Ambiente para la paz.* Bogotá: MMA, Cormagdalena, 1998.

Ramírez, María Clemencia. "La cuestión indígena colombiana (siglos XVI hasta el presente)", Bogotá: Manuscript, 1994.

———. "Los movimientos cívicos como movimientos sociales en el Putumayo: el poder visible de la sociedad civil y la construcción de una nueva ciudadanía." In *Movimientos sociales, Estado y democracia en Colombia,* edited by Mauricio Archila, and Mauricio Pardo. Bogotá: ICANH, CES, Universidad Nacional de Colombia, 2001.

Ramos, Alcida. *Indigenism: Ethnic Politics in Brazil.* Madison: University of Wisconsin Press, 1998.

———. "Cutting Through State and Class: Sources and Strategies of Self-representation in Latin America." In *Indigenous Movements, Self-representation, and the State in Latin America,* edited by Kay Warren, and Jean Jackson. Austin: University of Texas Press, 2002.

Reichel, Elizabeth. "La danta y el delfín: manejo ambiental e intercambio entre dueños de malocas y chamanes. El caso yukuna-matapí (Amazonas)." *Revista de Antropología* 5, no. 1–2 (1989).

Reichel-Dolmatoff, Gerardo. "A Brief Report on Urgent Ethnological Research in the Vaupés Area, Colombia, South America." *Bulletin of the International Committee on Urgent Anthropological and Ethnological Research* (1967).

Reichel-Dolmatoff, Gerardo. *Indios de Colombia.* Bogotá: Villegas Editores, 1991.

———. *Desana. Simbolismo de los indios tukano del Vaupés.* Bogotá: Departamento de Antropología de la Universidad de los Andes, 1968.

———. "Cosmology as Ecological Analysis. A View from the Forest." *Man.*11 (1976).

———. "Desana Animal Categories, Food Restrictions, and the Concept of Color Energies." *Journal of Latin American Lores* 4, no. 2 (1978).

———. *Los kogui.* Bogotá: Colcultura, vols. I–II, 1985.

———. *Goldwork and Shamanism: An Iconographic Study of the Golden Museum.* Medellín: Editorial Colina, 1988.

———. *The Forest Within. The World-View of the Tukano Amazonian Indians.* London: Themis Books, 1996.

Rodríguez, Gloria Amparo. *La consulta previa a pueblos indígenas y comunidades negras en el proceso para el otorgamiento de licencias ambientales.* Tesis de maestría en medio ambiente y desarrollo, Bogotá: Universidad Nacional de Colombia, 2001.

Rojas-Mix, Miguel. *América imaginaria.* Barcelona: Sociedad Estatal Quinto Centenario, Editorial Lumen, 1992.

Roldán, Roque. "Territorio étnicos, conflictos ambientales y paz." *Ambiente para la paz.* Bogotá: MMA, Cormagdalena, 1998.

Rootes, Chris. *Environmental Movements.* London: Frank Cass & Co., 1999.

Rubio, Rocío. *Gonawindúa Tayrona. Una organización indígena de la Sierra Neva-*

da de Santa Marta. Tesis de antropología. Bogotá: Universidad de los Andes, 1997.

Rutherford, Paul. "Ecological Modernization and Environmental Risk." In *Discourses of the Environment,* edited by Éric Danier. Oxford: Blackwell Publishers, 1999.

Sachs, Wolfgang. *Planet Dialectics. Exploration in Environment & Development.* London: Zed Books, 1999.

Said, Eduard. *Orientalism.* New York: Vintage Books, 1978.

———. *Covering Islam.* New York: Vintage Books, 1997.

Santos, Boaventura de Sousa. *La globalización del derecho. Los nuevos caminos de la regulación y la emancipación.* Bogotá: Universidad Nacional de Colombia, Ilsa, 1998.

Scott, Alan. "General Theories of Social Movements: Functionalism and Marxism." *Ideology and the New Social Movements,* 1990.

Scott, Joan. "Multiculturalism and the Politics of Identity." In *The Identity in Question,* edited by Jonh Rajchman. London: Routledge, 1995.

Sethi, Harsh. "Survival and Democracy: Ecological Struggles in India." In *New Social Movements in the South,* edited by Ponna Wignaraja. London, New Jersey: Zed Books, 1993.

Shiva, Vandana. *Close to Home. Women Reconnect Ecology, Health and Development Worldwide.* Gabriola Island, British Columbia (Canada): New Society Publishers, 1994.

Sierra, María Teresa. "Lenguaje, prácticas jurídicas y derecho consuetudinario indígena." In *Entre la ley y la costumbre. El derecho consuetudinario indígena en América Latina,* edited by Rodolfo Stavenhagen, and Diego Iturralde. México: Instituto Indigenista Americanista, Instituto Interamericano de Derechos Humanos, 1990.

Slater, Candice. *Entangled Edens. Visions of the Amazon.* Berkeley: University of California Press, 2002.

Slater, David. "New Social Movements and Old Political Questions: Some Problems of Socialist Theory with Relation to Latin America." Congreso Internacional de Americanistas, Amsterdam: Manuscript, 1988.

Stavenhagen, Rodolfo. "Derecho consuetudinario indígena en América Latina." In *Entre la ley y la costumbre. El derecho consuetudinario indígena en América Latina,* edited by Rodolfo Stavenhagen, and Diego Iturralde. México: Instituto Indigenista Americanista, Instituto Interamericano de Derechos Humanos, 1990.

Steiner, Christopher. "Travel Engravings and the Construction of the Primitive." In *Prehistories of the Future: The Primitivist Project and the Culture of Modernism,* edited by Elazar Barkan, and Ronald Bushb. Stanford: Stanford University Press, 1995.

Stone, Rosanne Allucquere. *Desire and Technology at the Close of Mechanical Age.* Cambridge: MIT Press, 1996.

Strathern, Marilyn. *Reproducing the Future: Anthropology, Kinship and the New Reproductive Technologies.* Manchester: Manchester University Press, 1992.

Sturgeon, Noel. "The Nature of Race. Discourses of Racial Difference in Ecofeminism." In *Ecofeminism. Women, Culture, Nature,* edited by Karen J. Warren. Bloomington, Indianapolis: Indiana University Press, 1997.

Tarrow, Sidney. *Power in Movement.* Cambridge: Cambridge University Press, 1998.
Taussing, Michael. *Shamanism, Colonialism, and the Wild Man.* Chicago: The University of Chicago Press, 1987.
Taylor, Charles. *Multiculturalism and the Politics of Recognition.* Princeton: Princeton University Press. 1992.
Tennant, Chris. "Indigenous Peoples, International Institutions, and the International Legal Literature from 1945–1993." *Human Rights Quarterly* 16 (1994).
Tilley, Virginia. "New Help or New Hegemony? The Transnational Indigenous Peoples' Movements and 'Being Indian' in El Salvador." *J. Latin Am. Stud.* 34 (2002).
Todorov, Tzvetan. *The Conquest of America. The Question of the Other.* New York: Harper and Row, 1984.
Torgovnick, Marianna. *Gone Primitive. Savage Intellects, Modern Lives.* Chicago, London: The University of Chicago Press, 1990.
Touraine, Alain. "An Introduction to the Study of Social Movements." *Social Research* 52, no. 4 (1985).
Turner, Terence. "An Indigenous People's Struggle for Socially Equitable and Environmentally Sustainable Production: The Kayapo Revolt Against Extractivism." *Journal of Latin American Anthropology* 9, no.1 (1995).
———. "Representation, Polyphony, and the Construction of Power." In *Indigenous Movements, Self-Representation, and the State in Latin America,* edited by Kay Warren, and Jean Jackson. Austin: University of Texas Press, 2002.
UAESPNN (Unidad Administrativa Especial del Sistema de Parques Nacionales Naturales). *Parques con la gente. Política de participación social en la conservación.* Bogotá: MMA, 2002.
Ulloa, Astrid. "Manejo tradicional de la fauna en procesos de cambio. Los embera en el Parque Nacional Natural Utría." In *Investigación y manejo de fauna para la construcción de sistemas sostenibles.* Cali: CIPAV, 1996.
———. "El nativo ecológico. Movimientos indígenas y medio ambiente en Colombia." In *Movimientos sociales, Estado y democracia en Colombia,* edited by Mauricio Archila, and Mauricio Pardo. Bogotá: ICANH, CES, Universidad Nacional de Colombia, 2001.
———. "Transformaciones en las investigaciones antropológicas sobre naturaleza, ecología y medio ambiente." *Revista Colombiana de Antropología* 37 (2001(a)).
———, ed. *Rostros culturales de la fauna. Las relaciones entre los humanos y los animales en el contexto colombiano.* Bogotá: ICANH, Fundación Natura, 2002(a).
———. "De una naturaleza dual a la proliferación de sentido: la discusión antropológica en torno a la naturaleza, la ecología y el medio ambiente." In *Repensando la naturaleza. Encuentros y desencuentros disciplinarios en torno a lo ambiental,* edited by Germán Palacio, and Astrid Ulloa. Bogotá: Universidad Nacional de Colombia-Leticia, Imani, ICANH, Colciencias, 2002(b).
———. "Pensando verde: el surgimiento y desarrollo de la conciencia ambiental global." In *Repensando la naturaleza. Encuentros y desencuentros disciplinarios en torno a lo ambiental,* edited by Germán Palacio, and Astrid Ulloa. Bogotá: Universidad Nacional de Colombia-Leticia, Imani, ICANH, Colciencias, 2002(c).

Ulloa, Astrid, Heidi Rubio-Torgler, and Claudia Campos. "Conceptos y metodologías para la preseleccíon y análisis de alternativas de manejo de fauna de caza con indígenas embera en el Parque Nacional Natural Utría." In *Manejo de fauna con comunidades rurales,* edited by Claudia Campos, Astrid Ulloa, and Heidi Rubio-Torgler. Bogotá: ICANH, Fundación Natura, Orewa, OEI, MMA, 1996.

Umiyac (Unión de Médicos Indígenas Yageceros de Colombia). *Encuentros de taitas en la Amazonía colombiana. Ceremonias y reflexiones.* Bogotá: UMIYAC, 1999.

Urban, Greg, and Joel Sherzer, ed. *Nation-States and Indians in Latin America.* Austin: University of Texas Press, 1991.

Uribe, Carlos. A. "De la Sierra Nevada de Santa Marta, sus ecosistemas, indígenas y antropólogos." *Revista de Antropología* 4, no.1 (1988).

———. "La etnografía de la Sierra Nevada de Santa Marta y las tierras bajas adyacentes." In *Geografía Humana de Colombia. Nordeste Indígena.* Bogotá: Instituto de Cultura Hispánica, tomo II, 1993.

Van Cott, Donna Lee (ed.). *Indigenous Peoples and Democracy in Latin America.* New York: St. Martin's Press, 1994.

———. *The Friendly Liquidation of the Past. The Politics of Diversity in Latin America.* Pittsburgh: University of Pittsburgh Press, 2000.

Van der Hammen, María Clara. *El manejo del mundo. Naturaleza y sociedad entre los yukuna de la Amazonía colombiana.* Bogotá: Tropenbos, 1992.

Varese, Stefano. "Pueblos indígenas y globalización en el umbral del tercer milenio." In *Articulación de la diversidad. Pluralidad étnica, autonomía y democratización en América Latina, Grupo de Barbados,* comp.: Georg Grunberg. Quito: Biblioteca Abya-Yala 27, 1995.

———, comp.: *Pueblos indios, soberanía y globalismo.* Quito: Ediciones Abya-Yala, no.32, 1996.

———. "Parroquialismo y globalización. Las etnicidades indígenas ante el tercer milenio." In *Pueblos indios, soberanía y globalismo,* comp.: Stefano Varese. Quito: Ediciones Abya-Yala, no.32, 1996(a).

———. "The New Environmentalist Movement of Latin American Indigenous People." In *Valuing Local Knowledge. Indigenous People and Intellectual Property Rights,* edited by Stephen B. Brush, and Doreen Stabinsky. Washington: Island Press, 1996(b).

Vasco, Luis Guillermo. *Entre selva y páramo, viviendo y pensando la lucha india.* Bogotá: ICANH, 2002.

Villegas, Marta. *Los wiwa: nociones de equilibrio y movimiento.* Tesis de pregrado en antropología, Universidad de los Andes. Bogotá: Manuscript, 1999.

Vivanco, Luis. "Seeing Green: Knowing and Saving the Environment on Film." *American Anthropologist* 104, no.4 (2002).

Wade, Peter. *Race and Ethnicity in Latin America.* London: Pluto Press, 1997.

———. "Entre la homogeneidad y la diversidad." In *Antropología en la modernidad,* edited by María Victoria Uribe, and Eduardo Restrepo. Bogotá: Ican, 1997(a).

Wapner, Paul. "Politics Beyond the State: Environmental Activism and World Civic Politics." *World Politics* 47, no.3 (1994).

———. "Environmental Activism and Global Civil Society." *Dissent* 41 (1995).

Warren, Kay B. "Indigenous Movements as a Challenge to the Unified Social Movement Paradigm for Guatemala." ." In *Cultures of Politics, Politics of Cultures. Re-visioning Latin America Social Movements,* edited by Sonia Alvarez, Evelina Dagnino and Arturo Escobar. Boulder: Westview Press, 1998.

———. *Indigenous Movements and Their Critics.* Princeton: Princeton University Press, 1998(a).

Warren, Kay, and Jackson, Jean, ed. *Indigenous Movements, Self-Representation, and the State in Latin America.* Austin: University of Texas Press, 2002.

Watts, Rob. "Government and Modernity: An Essay in Thinking Governmentality." *Arena Journal* 2 (1993–1994).

Webb, Virginia-Lee. "Manipulated Images." In *Prehistories of the Future,* edited by Elazar Barkan, and Roland Bush. Stanford: Stanford University, 1995.

Williams, Bruce A., and Albert R. Mathenny. *Democracy, Dialogue, and Environmental Disputes. The Contested Languages of Social Regulations.* New Haven, London: Yale University Press, 1995.

Wilmer, Franke. *The Indigenous Voice in World Politics.* London: Sage Publications, 1993.

Yashar, Deborah. "Indigenous Protest and Democracy in Latin America." In *Constructing Democratic Governance: Latin America and the Caribbean,* edited by Jorge I. Domínguez, and Abraham F. Lowenthal. London: Johns Hopkins University Press, 1996.

———. "Contesting Citizenship. Indigenous Movements and Democracy in Latin America." *Comparative Politics,* October (1998).

———. "The Post-liberal Challenge in Latin America." Annual American Political Science Association Conference. Boston: Manuscript, 1998(a).

———. "Democracy, Indigenous Movements, and the Postliberal Challenge in Latin America." *World Politics.* 52, no.1 (1999).

Young, Zoe. "After the Greenrush. Saving Nature for Capital with the Global Environmental Facility." In http//www.newgreenorder.info, 2002.

Zambrano, Carlos. "Conflictos y cambios en el proceso de modernización del macizo colombiano. Un caso de alteridad étnica." In *Modernidad, identidad y desarrollo,* edited by María Lucía Sotomayor. Bogotá: ICAN, 1998.

———. "Un análisis del movimiento social y étnico del macizo Colombiano." In *Movimientos sociales, Estado y democracia en Colombia,* edited by Mauricio Archila, and Mauricio Pardo. Bogotá: ICANH, CES, Universidad Nacional de Colombia, 2001.

Index

For Product Safety Concerns and Information please contact our EU representative GPSR@taylorandfrancis.com
Taylor & Francis Verlag GmbH, Kaufingerstraße 24, 80331 München, Germany

www.ingramcontent.com/pod-product-compliance
Ingram Content Group UK Ltd.
Pitfield, Milton Keynes, MK11 3LW, UK
UKHW020937180425
457613UK00019B/444

9780415884051